Basic English Skills for Beginners
Majored in Chemistry and Chemical Engineering

化学化工专业英语入门

翁哲慧　葛丹丹　王宝玲　编著

化学工业出版社

·北京·

内容简介

《化学化工专业英语入门》对化学实验室安全相关知识、实验记录要领以及普通化学化工实验仪器等一般共性知识进行了阐述，同时介绍了化学化工学科的分类、化学化工学术交流的相关英语知识等内容。从英文视角介绍从化学化工学术基础再到化学化工学术交流等相关知识，层层递进而又互相关联——从系统阐释传统化学化工专业基础知识，逐渐转向对更高层次的化学化工类学术交流活动的介绍。每个单元的文章后都有课文难/长句的注释，很多文章附有图片说明，以增加文章的生动性和可读性。

本书可作为普通高等院校化学、化工、材料及相关专业本科生和研究生的专业英语教材，同时也供化学、化工、材料领域的科研和工程技术人员参考。

图书在版编目（CIP）数据

化学化工专业英语入门 / 翁哲慧，葛丹丹，王宝玲编著． -- 北京：化学工业出版社，2024．8． -- ISBN 978-7-122-46139-1

Ⅰ．O6；TQ

中国国家版本馆 CIP 数据核字第 2024J4E729 号

责任编辑：张　艳　　　　　　　　文字编辑：曹　敏
责任校对：宋　玮　　　　　　　　装帧设计：王晓宇

出版发行：化学工业出版社
　　　　　（北京市东城区青年湖南街 13 号　邮政编码 100011）
印　　装：北京科印技术咨询服务有限公司数码印刷分部
787mm×1092mm　1/16　印张 15¼　彩插 1　字数 393 千字
2024 年 8 月北京第 1 版第 1 次印刷

购书咨询：010-64518888
售后服务：010-64518899
网　　址：http://www.cip.com.cn

凡购买本书，如有缺损质量问题，本社销售中心负责调换。

定　　价：49.80 元　　　　　　　　　　　　　版权所有　违者必究

前言

英语作为广泛使用的一种语言,在国际交流中起着十分重要的作用,也是我国大多数学生学习的第二种语言。本书是为化学化工类及相关专业的学生在修完大学公共英语之后,进入专业英语学习而编写的,目的是让学生了解英语在相关专业中的作用,并且掌握化学化工专业英语的语言规律,以便于理解科技英语文献、扩大英语词汇量、巩固英语知识,进而提高学生应用英语的能力。

目前出版的大部分化学化工专业英语的教材的受众群体是有着过硬的化学化工专业背景以及对于科技英语有较好的基础的一类人员,这类教材的专业面广、内容深、针对性强。而对于科技英语基础较为薄弱的群体来说,现阶段的大多数化学化工专业英语教材难度偏大,对于诸如化学分子式与化学方程式的读/写方法、化合物结构和形态的英文描述、实验室工作相关英文记录、普通化学化工类国际学术交流等基础性专业英语知识还缺乏完整的介绍。因此,教师在为科技英语基础较为薄弱的学生讲授"化学化工专业英语"课程时,存在参考教材与学生水平脱节的问题。为此,我们特别编写了这本专门针对基础英语掌握程度一般的化学化工类及其相关专业的本科生或研究生均适用的入门级专业英语的教材。

本书共分为三个部分,Part A 共 2 个单元,主要是宏观介绍化学化工学科的特点,内容涉及化学实验室安全相关知识、实验记录要领以及普通化学化工实验仪器等一般共性的知识,尤其注重的是基础性知识和广泛性科技英语词汇;Part B 共 7 个单元,主要包括化学化工学科的分类介绍,涵盖化学化工初级知识以及基础化工相关的专业英语,如化合物的命名及化工过程单元操作等;Part C 包含 2 个单元,主要涵盖化学化工学术交流的相关英语知识,如:参加学术活动相关的英语书面/口头交流的基础知识等。以上三大部分,用英文视角介绍从化学化工学术基础再到化学化工学术交流等相关知识,层层递进而又互相关联——从系统阐释传统化学化工专业基础知识,逐渐转向对更高层次的化学化工类学术交流活动的介绍。希望化学化工专业的学生通过学习本书内容,可以具备进一步学习国际上更为先进的化学化工相关科学知识的基础;同时,通过本教材的编写,我们试图改变当今大学生仅将英语学习局限于普通英文的"阅读理解",而完全忽视学术交流的需求这一现状。在编写本书时,我们有意避免对传统美国式主流专有词汇的直接引用,相反,我们结合目前中国相关词汇的英文表达,将这些内容适当地展现出来,并且有意识地增加了许多中国元素;此外,本书的编写还适当使用了当今中国化学化工发展过程中最新的信息,如"碳达峰""碳中和"以及与化学化工相关的中国特色的词汇等。本书在吸收国内同类教材优点的同时,根据现代科技的发展补充了一些新知识,如环境与能源、国内外化学工程师的职业规划、化学相关信息的来源等内容。

本教材共计 11 个单元的课文,每个单元的文章后都有课文难/长句的注释,很多文章附有图片说明,以增加文章的生动性和可读性。本书内容涉及宽泛,词汇量较广,力求深入浅出,对部分专业词汇注有音标和构词规律说明。为了帮助读者更好地掌握这些细节知识，我们在每个单元的最后都设有"Trivial — Last but not the Least"（细节知识点）部分，涵盖了从读写化学分子式/方程式到正确使用英文表达化学用语、化学计量单位等化学工作者接触过但可能难以牢固掌握的细节知识。本书适合作为普通高等院校化学、化工、材料及相关专业本科生和研究生的专业英语教材，同时也供化学、化工、材料领域的科研和工程技术人员参考。我们希望读者通过阅读本书可以更全面地掌握化学化工英语入门知识，为未来的学习和工作打下坚实的基础。

  本书是由昆明学院化学化工学院的翁哲慧、葛丹丹和王宝玲三位老师共同编著。在获取英文资料和描述化学问题方面，得到了常州大学的张致慧、云南民族大学的赵梦楠、美国阿拉巴马大学亨茨维尔分校的凌颉以及昆明学院的鞠海东、乔永锋、吴琼、周小华、李湘广、陈人杰、杨波、杨艳华和赵娜等众多老师的精心指导和无私帮助，对此我们深表感激。另外，在校对书稿的过程中，我们也得到了昆明学院材料与化工专业研究生黄仙金、江小舟、李琦、陈充玉等同学的鼎力支持，对此我们表示衷心的感谢。

  尽管几经修改，力求完善此书，但由于作者英语和专业水平的限制，加之时间紧迫，难免存在不足之处，诚恳地希望各位读者、同行能够提出宝贵的批评意见和建议，以便我们在修订时进行改进。

翁哲慧
2024 年 5 月于昆明

Contents

Part A Basic Understanding of Chemistry ··· 001

Unit 1 Overview of Chemistry ··· 002
1.1 What is Chemistry? ··· 002
1.2 History of Chemistry ··· 002
1.3 Chemistry in the Twenty-first Century ··· 003
1.4 Chemists and Chemical Engineers ··· 005
1.5 Branches of Chemistry ··· 005
1.6 Chemical Reactions ··· 006
1.7 Types of Chemical Reactions ··· 007
1.8 The Study of Chemistry ··· 009
1.9 Scientific Method ··· 010
New Words and Expressions ··· 011
Notes ··· 013
Trivial — Last but not the Least ··· 014

Unit 2 Working in the Lab ··· 016
2.1 Safety in the Lab ··· 016
2.2 Disposal of Chemical Waste ··· 017
2.3 How to Deal with Bad Lab Partners? ··· 017
2.4 The Most Common Injuries in a Chemistry Lab ··· 019
2.5 The Lab Notebook for Chemistry Lab Report ··· 020
2.6 Basic Laboratory Apparatus and Manipulation ··· 022
    2.6.1 Apparatus ··· 022
    2.6.2 Manipulation ··· 027
New Words and Expressions ··· 029
Notes ··· 032
Trivial — Last but not the Least ··· 032

Part B Introduction to Fundamentals of Chemistry and Chemical Engineering ··· 035

Unit 3 Basic Information of Inorganic Chemistry ··· 036
3.1 Overview ··· 036

3.2　Four Types of Inorganic Chemical Reactions ··037
3.3　The Important Inorganic Elements — Actinide ······································038
New Words and Expressions ···039
Notes ··040
Trivial — Last but not the Least ···041

Unit 4　Rules of Inorganic Nomenclature ··044

4.1　Oxidation Number ··044
4.2　Coordination Number ··044
4.3　Use of Multiplicative Prefixes, Enclosing Marks, Numbers, and Letters ············045
 4.3.1　Multiplicative Prefixes ···045
 4.3.2　Enclosing Marks ···045
 4.3.3　Numbers ···046
 4.3.4　Italic Letters ··046
4.4　Elements ··046
4.5　Formulae and Names of Compounds in General ··047
4.6　Names for Ions and Radicals ···049
4.7　Inorganic Acids ··051
 4.7.1　Oxoacid ··051
 4.7.2　Acid Anions ···052
 4.7.3　Binary Hydrogen Compounds (Binary Acid) ···053
4.8　Salts ··053
 4.8.1　Simple Salts ···053
 4.8.2　Salts Containing Acid Hydrogen ··053
 4.8.3　Hydrated Salts ···053
4.9　Coordination Compounds ···054
 4.9.1　Indication of Oxidation Number and Proportion of Constituents ·························054
 4.9.2　The Names of Anionic Ligands, whether Inorganic or Organic, End in -o ············055
 4.9.3　Alternative Modes of Linkage of Some Ligands ···056
 4.9.4　Complexes with Unsaturated Molecules or Groups ··057
 4.9.5　Compounds with Bridging Atoms or Groups ···057
 4.9.6　Di- and Polynuclear Compounds without Bridging Groups; Direct Linkage between Coordination Centers ···057
 4.9.7　Homoatomic Aggregates ··058
4.10　Addition Compounds ···058
4.11　Prefixes or Affixes Used in Inorganic Nomenclature ·······································059
New Words and Expressions ···060
Notes ··065
Trivial — Last but not the Least ···066

Unit 5　A Short Introduction to Organic Chemistry ····························067

5.1　Overview ··067
5.2　Types of Organic Chemical Reactions ··069

5.3 Name Reactions in Organic Chemistry ... 072

- 5.3.1 Acetoacetic-Ester Condensation Reaction ... 073
- 5.3.2 Acetoacetic Ester Synthesis ... 073
- 5.3.3 Acyloin Condensation ... 073
- 5.3.4 Alder-Ene Reaction or Ene Reaction ... 073
- 5.3.5 Aldol Reaction or Aldol Addition ... 074
- 5.3.6 Aldol Condensation Reaction ... 074
- 5.3.7 Appel Reaction ... 074
- 5.3.8 Arbuzov Reaction or Michaelis-Arbuzov Reaction ... 074
- 5.3.9 Arndt-Eistert Synthesis Reaction ... 074
- 5.3.10 Azo Coupling Reaction ... 075
- 5.3.11 Baeyer-Villiger Oxidation Reactions ... 075
- 5.3.12 Baker-Venkataraman Rearrangement ... 075
- 5.3.13 Balz-Schiemann Reaction ... 075
- 5.3.14 Bamford-Stevens Reaction ... 075
- 5.3.15 Barton Decarboxylation ... 076
- 5.3.16 Barton Deoxygenation Reaction: Barton-McCombie Reaction ... 076
- 5.3.17 Baylis-Hillman Reaction ... 076
- 5.3.18 Beckmann Rearrangement Reaction ... 076
- 5.3.19 Benzilic Acid Rearrangement ... 076
- 5.3.20 Benzoin Condensation Reaction ... 077
- 5.3.21 Bergman Cycloaromatization—Bergman Cyclization ... 077
- 5.3.22 Bestmann-Ohira Reagent Reaction ... 077
- 5.3.23 Biginelli Reaction ... 077
- 5.3.24 Birch Reduction Reaction ... 077
- 5.3.25 Bicschler-Napieralski Reaction — Bicschler-Napieralski Cyclization ... 077
- 5.3.26 Blaise Reaction ... 078
- 5.3.27 Blanc Reaction ... 078
- 5.3.28 Bohlmann-Rahtz Pyridine Synthesis ... 078
- 5.3.29 Bouveault-Blanc Reduction ... 078
- 5.3.30 Brook Rearrangement ... 078
- 5.3.31 Brown Hydroboration ... 078
- 5.3.32 Bucherer-Bergs Reaction ... 079
- 5.3.33 Buchwald-Hartwig Cross Coupling Reaction ... 079
- 5.3.34 Cadiot-Chodkiewicz Coupling Reaction ... 079
- 5.3.35 Cannizzaro Reaction ... 079
- 5.3.36 Chan-Lam Coupling Reaction ... 079
- 5.3.37 Crossed Cannizzaro Reaction ... 080
- 5.3.38 Friedel-Crafts Reaction ... 080
- 5.3.39 Huisgen Azide-Alkyne Cycloaddition Reaction ... 080
- 5.3.40 Itsuno-Corey Reduction — Corey-Bakshi-Shibata Reduction ... 080
- 5.3.41 Seyferth-Gilbert Homologation Reaction ... 081

New Words and Expressions ·· 081
Notes ··· 084
Trivial — Last but not the Least ·· 085

Unit 6 Rules of Nomenclature of Organic Chemistry ································ 087

6.1   Hydrocarbons ·· 087
  6.1.1   Alkanes ·· 087
  6.1.2   Alkenes ·· 090
  6.1.3   Alkynes ·· 092
6.2   Heteroatom Functional Groups ·· 092
  6.2.1   Alcohols, Phenol, and Thiols ·· 093
  6.2.2   Ethers ·· 095
  6.2.3   Amines ··· 095
  6.2.4   Organohalogen Compounds ·· 097
  6.2.5   Aldehydes and Ketones ··· 098
  6.2.6   Carboxylic Acid ·· 098
  6.2.7   Carboxylic Acid Derivatives ··· 099
New Words and Expressions ·· 101
Notes ··· 103
Trivial — Last but not the Least ·· 103

Unit 7 Introduction to Analytical Chemistry ·· 107

7.1   Overview ··· 107
7.2   The Scope of Analytical Chemistry ··· 108
7.3   The Functions and Patterns of Analytical Chemistry ···························· 109
  7.3.1   Functions of Analytical Chemistry ·· 109
  7.3.2   General Patterns of Analytical Chemistry ··································· 110
7.4   Analytical Methods ·· 111
  7.4.1   Types of Analytical Methods ·· 111
  7.4.2   Trends in Analytical Methods and Procedures ······························ 112
7.5   Some Modern Methods of Analytical Chemistry ································ 113
  7.5.1   High Performance Liquid Chromatography ································· 113
  7.5.2   High Performance Capillary Electrophoresis ································ 116
  7.5.3   Infrared Spectrophotometry ·· 118
  7.5.4   Nuclear Magnetic Resonance ·· 119
  7.5.5   Atomic Absorption ·· 121
New Words and Expressions ·· 124
Notes ··· 127
Trivial — Last but not the Least ·· 128

Unit 8 A Map of Physical Chemistry ·· 129

8.1   Overview ··· 129
  8.1.1   What Is Physical Chemistry? ·· 129

  8.1.2 Theories of Physical Chemistry ·· 130
  8.1.3 What Do Physical Chemists Do? ·· 131
  8.1.4 Perspective of Physical Chemistry ·· 131
 8.2 Briefing on Thermodynamics ··· 132
 8.3 Basic Concepts of Thermodynamics ··· 133
 8.4 Introduction to Chemical Equilibrium and Kinetic Theory ····················· 135
  8.4.1 Equilibrium ··· 135
  8.4.2 Chemical Kinetics ·· 139
 New Words and Expressions ··· 140
 Notes ··· 142
 Trivial — Last but not the Least ·· 143

Unit 9 Chemical Engineering ·· 146

 9.1 Overview ·· 146
 9.2 The Difference between Chemistry and Chemical Engineering ··············· 147
 9.3 Future Opportunities in Chemical Engineering ···································· 148
 9.4 Introduction to Basic Chemical Engineering Concepts ·························· 149
  9.4.1 Heat Transfer and Heat Exchangers ·· 149
  9.4.2 The Material Balance and the Energy Balance ······························ 150
  9.4.3 The Ideal Contact and the Rates of an Operation ·························· 150
  9.4.4 Chemical Engineering Process Related to Solids ·························· 151
  9.4.5 Fluidization ··· 153
  9.4.6 Supercritical Fluids and Supercritical Fluid Extraction ··················· 154
  9.4.7 Extraction and Liquid-Liquid Extraction ····································· 156
  9.4.8 Evaporation and Crystallization ··· 157
  9.4.9 Dust Removal and Centrifugal Settling Process ···························· 158
  9.4.10 Batch and Continuous Process ·· 159
  9.4.11 Chemical Manufacturing Process ··· 159
 New Words and Expressions ··· 160
 Notes ··· 163
 Trivial — Last but not the Least ·· 164

Part C English for Basic Academic Communication ···················· 167

Unit 10 International Conference ··· 168

 10.1 An Overview of Academic Communication ······································· 168
 10.2 International Conference ·· 169
  10.2.1 Sources for Conference Information ·· 169
  10.2.2 Conference Organizers and Session Modes ································ 171
  10.2.3 Papers, Abstracts, Posters and Proposals Submitted to Conference ···· 174
  10.2.4 Financial Assistance for Attending Meetings ······························ 180
  10.2.5 Rehearsal, Attendance and Culture Considerations ······················· 181

New Words and Expressions ·· 187
　　Notes ··· 193
　　Trivial — Last but not the Least ·· 193
Unit 11　Letters for Academic Communication ································· 194
　11.1　Introduction to Letter Writing ·· 194
　　11.1.1　Basic Principles for Letter Writing ···································· 194
　　11.1.2　Styles of Punctuation and Format ······································ 195
　　11.1.3　Necessary Parts of a Formal Letter ···································· 195
　　11.1.4　Optional Parts of a Letter ·· 199
　　11.1.5　Stationery and Envelope ·· 201
　　11.1.6　Sample letter with Common Formats ···································· 202
　11.2　Introduction to Email Writing ··· 202
　　11.2.1　Email Format ·· 203
　　11.2.2　Subject Heading ·· 203
　　11.2.3　Salutation ··· 203
　　11.2.4　The Body ·· 203
　　11.2.5　Previous Message and Quotes ·· 204
　　11.2.6　Closing ··· 205
　　11.2.7　Signature ··· 206
　　11.2.8　File Attachments ··· 206
　　11.2.9　Reread and Save before Clicking the "Send" Button ···················· 206
　11.3　Examples of Letters in Different Situations ································· 207
　　11.3.1　Examples of Letters Relating to Conference ··························· 207
　　11.3.2　Letters for Academic Visits and Cooperation ·························· 209
　　11.3.3　Letters for Research Positions ·· 211
　　New Words and Expressions ·· 212
　　Notes ·· 214
　　Trivial — Last but not the Least ·· 215

Appendix ··· 217

　A1　IUPAC Names and Symbols of Elements ······································ 218
　A2　General Stems and Affixes ··· 221
　A3　Common English Terms Used in the Laboratory ······························ 232

References ·· 234

Part A
Basic Understanding of Chemistry

Unit 1
Overview of Chemistry

1.1　What is Chemistry?

　　Chemistry, like physics and biology, is a natural science. In fact, there is considerable overlap between chemistry and these other disciplines. Its focus lies on the study of matter, including atoms, compounds, chemical reactions, and chemical bonds. Chemists explore the properties of matter, its structure, and how it interacts with other matter. Chemistry is the branch of science that examines matter — anything with mass and volume, and how it transforms when exposed to varying energy, environments, and conditions. In other words, chemists try to understand the unchanging properties of matter, as well as why and how some of the properties of matter change. Everything you see, hear, or smell involves chemistry. Cooking dinner, washing clothes, experiencing the effects of caffeine, and making ice cubes are just a few of the millions of examples of ways that people use chemistry in their everyday lives. Another important facet of chemistry is that chemists seek to use the knowledge they have accumulated to try to create new matter and improve the chemicals and materials that are available. For example: research to develop new medicines is chemistry; the search for better sustainable energy relies on chemistry; the development of new smells and tastes is a form of chemistry.

1.2　History of Chemistry

　　Chemistry has existed as long as humans have, but it was not always formalized as a scientific field as it is today. During prehistoric times, humans documented the discovery of metals and other materials. Aristotle, the ancient Greek philosopher, posited the existence of four basic elements: water, air, earth, and fire. While his theory was incomplete, it reflects humanity's ongoing desire to comprehend the world around us.

　　Alchemy, which emerged in the 8th century, was also a significant precursor to modern chemistry wherein alchemists endeavored to transmute cheap metals into gold and create elixirs for life through chemical reactions. Although alchemists lacked an understanding of how many processes worked, they were very thorough and recorded information from their experiments. At the same time, the alchemist Jābir ibn Ḥayyān wrote many scrolls detailing methods of distillation, crystallization, sublimation, and evaporation and developed an early classification system for chemicals based on their properties.

　　Traditional chemistry experienced a great breakthrough in the 17th century when Robert

Boyle introduced the scientific method. Formal experimentation and the repetition of controlled studies facilitated significant advancements in human comprehension of the world during the 17th and 18th centuries.

1.3   Chemistry in the Twenty-first Century

Chemistry is the study of matter and the changes it undergoes. Chemistry is often called the central science, because a basic knowledge of chemistry is essential for students of biology, physics, geology, ecology, and many other subjects. Indeed, it is central to our way of life; without it, we would live shorter lives in what we would consider primitive conditions, without cars, electricity, computers, CDs, and many other everyday conveniences.

Although chemistry is an ancient science, and formal experimentation and the repetition of controlled studies enabled leaps in the human understanding of the world in the 17th and 18th centuries, it was not until the 19th century that chemistry became a practical science. For example, Galileo's telescope, the paradigmatic instrument of discovery in pure science, emerged from the entirely pragmatic tradition of making lenses for spectacles. Similarly, the modern foundations of chemistry were laid in the nineteenth century, when intellectual and technological advances enabled scientists to break down substances into smaller and smaller components, explaining many of their physical and chemical properties. The rapid development of increasingly sophisticated technology throughout the twentieth century has given us even greater means of studying things that cannot be seen with the naked eye[1]. For example, using computers and special microscopes, chemists can analyze the structure of atoms and molecules — the basic units on which the study of chemistry is based — and design new substances with specific properties, such as medicines and environmentally friendly consumer products.

As China entered the 21st century, it is highly likely that the chemical industry will play a significant role in science and technology to achieve the "dual carbon" goal, the assertion based on the Chinese government's continued commitment to achieve "carbon peak" and "carbon neutrality" in an effort to combat global climate change[2]. Before delving into the details of substances and their transformations, let's take a look at some of the cutting-edge areas currently being explored by chemists, including **medicinal chemistry, energy and the environment, materials and technology**, and **food and agriculture**. A deeper understanding of this discipline can enhance appreciation of its impact on society and individuals.

Medicinal chemistry is a subject based on chemistry and biology that studies the structure and activity of drugs. Medicinal chemistry is a comprehensive discipline that discovers and invents new drugs, synthesizes chemical drugs, elucidates the nature of medicinal chemistry and studies the law of interaction between drug molecules and body cells (biological macromolecules). It is an important and leading discipline in the pharmaceutical field and has made significant contributions to human health. Chemists in the pharmaceutical industry are researching powerful drugs with few or no side effects to treat cancer, AIDS and many other diseases, as well as drugs to increase the number of successful organ transplants. More broadly, a better understanding of the mechanisms of ageing will lead to longer and healthier lives for the world's population.

Energy and the environment are two emerging issues in the field of chemistry. Energy is a by-product of many chemical processes, and as the demand for energy continues to grow, chemists around the world are actively trying to find new sources of energy. Energy and the environment are two important aspects that are interrelated and influence each other. Chemistry, as a broad discipline, has important application value in areas such as materials, energy and the environment. In the field of energy, chemistry can help solve the problems of energy conversion and storage. For example, solar cells, fuel cells, photocatalytic materials, etc. all rely on chemical reactions to achieve energy conversion. In recent years, China has prioritized the development of the new energy industry. By the end of 2022, China's production and sales of new energy vehicles had already surpassed those of any other country. Chemistry plays a crucial role in energy science and environmental management. In the field of energy science, chemistry can be used to develop energy storage devices such as lithium-ion batteries and supercapacitors; in the field of environmental management, chemistry can help treat pollutants such as wastewater and exhaust gases to improve environmental quality. For example, artificially designed and synthesized chemical reagents can effectively degrade organic pollutants in wastewater, and chemical adsorption can also remove harmful gases from the atmosphere, improving air quality.

Materials and technology are the material basis of all human life and production activities. Materials, especially functional materials, have become increasingly important in today's industrial and information society. The discovery and practical application of a new material often changes the concepts and behavior of an industry or even people. The ability to research and develop new materials has become an important indicator of a country's scientific and technological level and its level of productivity development. By using chemical methods of self-assembly, polymer synthesis, and nanostructure preparation to control the structure and properties of materials, new materials with specific functions can be designed. For example, chemical methods can be used to prepare biomedical materials, photocatalytic materials, battery materials, and so on, and these materials have a wide range of applications. The discovery of new materials not only promotes the development of related scientific fields, but also creates and drives a number of high-tech industries through their applications. Therefore, the originality of materials science and the improvement of materials technology innovation capabilities are also important issues for the development of the discipline of chemistry.

Food and agriculture are the basis for the origin and development of chemistry. Chemistry is not only at the core of innovation and application in agriculture and food, it is also an important component of agricultural production and scientific experimentation. How will the world's rapidly growing population be fed? What are the various factors that impact agricultural production, such as soil fertility, crop-damaging pests and diseases, and nutrient-competing weeds? To meet the food demands of the twenty-first century, new and novel approaches to agriculture must be developed. It has been demonstrated that modern technology has the potential to cultivate crops that are larger and of higher quality through the utilization of genetic methods. These methods produce crops that enable plants to discourage pesky insects from reproducing. Techniques for improving agricultural production can be applied to various crops to increase yields and crop frequency. Chemists can also develop methods to produce fertilizers with reduced environmental impact and select chemicals to target specific weeds.

1.4 Chemists and Chemical Engineers

Although chemical engineers and chemists are closely related, it should be noted that they have distinct roles. Unlike chemists who experiment with various chemicals using test tubes or beakers, chemical engineers are professionals who specialize in independently designing, constructing, and producing chemical engineering expertise. Thus, chemical engineers serve as the bridge or converter between the chemical laboratory and the chemical plant.

Employers of chemists and chemical engineers include various government agencies, such as environmental protection agencies at all levels. Other employers include large pharmaceutical factories such as Sinopharm, state-owned steel and construction companies, large joint ventures such as Wuxi Pharmatech (Cayman) Inc., national private companies such as Huawei, BYD, and DJI Group, and giant multinational corporations such as Johnson & Johnson and DuPont.

1.5 Branches of Chemistry

Inorganic chemistry is the discipline that focuses on compounds beyond organic chemistry, specifically those that lack a C—H bond. Carbon is present in certain inorganic compounds, but the vast majority of inorganic compounds are made up of metals. Topics of interest to inorganic chemists include ionic compounds, organometallic compounds, minerals, cluster compounds, and solid state compounds.

Organic chemistry involves the scientific study of chemical processes occurring in living organisms and their molecules. An organic chemist might explore the organic reactions, the structure and properties of organic molecules, polymers, drugs, or fuels.

Analytical chemistry involves studying the chemistry of matter and creating tools to quantify its properties. Analytical chemistry includes quantitative and qualitative analysis, separations, extractions, distillation, spectrometry and spectroscopy, chromatography, and electrophoresis. Analytical chemists develop standards, chemical methods, and instrumental methods.

Physical chemistry is a branch of chemistry that applies physics to the study of chemistry, which commonly includes the applications of thermodynamics and quantum mechanics to chemistry.

Biochemistry is the study of the chemical processes that occur within living organisms. Examples of key molecules include proteins, nucleic acids, carbohydrates, lipids, drugs, and neurotransmitters. Sometimes, this discipline is considered a subdiscipline of organic chemistry. Biochemistry is closely related to molecular biology, cell biology, and genetics.

Other Branches of Chemistry

There are alternative methods of classifying chemistry branches that may include the above disciplines as a primary branch. Other examples of chemistry branches include:

Astrochemistry examines the abundance of elements and compounds in the universe, their reactions with each other, and the interaction between radiation and matter.

Chemical kinetics (or simply "kinetics") studies the rates of chemical reactions and processes and the factors that affect them.

Electrochemistry studies the transport of charge in chemical systems, focusing on electrons as the primary charge carrier. It also studies the behavior of ions and protons.

Green chemistry looks at ways of minimizing the environmental impact of chemical processes. This includes remediation as well as ways of improving processes to make them more eco-friendly.

Geochemistry examines the nature and properties of geological materials and processes.

Nuclear chemistry studies the reactions between protons, neutrons, and subatomic particles, while most forms of chemistry are mainly concerned with the interactions between electrons in atoms and molecules.

Polymer chemistry deals with the synthesis and properties of macromolecules and polymers.

Quantum chemistry applies quantum mechanics to model and explore chemical systems.

Radiochemistry explores the nature of radioisotopes, the effects of radiation on matter, and the synthesis of radioactive elements and compounds.

Theoretical chemistry is the branch of chemistry that applies mathematics, physics, and computer programming to answer chemistry questions.

Inorganic Chemistry Journals and Publications

There are numerous publications devoted to advances in inorganic chemistry. Journals include Inorganic Chemistry, Polyhedron, Journal of Inorganic Biochemistry, Dalton Transactions, and Bulletin of the Chemical Society of Japan[3].

1.6 Chemical Reactions

A **chemical reaction** is any transformation from one set of chemicals into another set. If the initial and final substances remain the same, a transformation might have taken place, but not a chemical reaction. A **chemical reaction** is a reaction involves the rearrangement of molecules or ions into a different structure. Contrast this with a physical change where the appearance changes but the molecular structure remains the same, or a nuclear reaction where the composition of the atomic nucleus undergoes a change. In a chemical reaction, the atomic nucleus remains unchanged, while electrons can be shared or transferred to create or break down chemical bonds[4]. In both physical changes and chemical changes (reactions), the quantity of atoms for each element remains constant both before and after a process occurs. In a physical change, the atoms preserve their original molecular and compound arrangement. However, during a chemical reaction, atoms combine to form new molecules and compounds, resulting in the creation of new products.

Signs of chemical reaction. Since you can't look at chemicals at a molecular level with the naked eye, it's helpful to know signs that indicate a reaction has occurred. Chemical reactions often cause temperature changes, bubbles, color changes, and/or precipitate formation.

Chemical reactions and chemical equations. The atoms and molecules that interact are called the reactants. The atoms and molecules produced by the reaction are called products. Chemists use a shorthand notation called a chemical equation to indicate the reactants and the products. In this notation, the reactants are listed on the left side, the products are listed on the right side, and the reactants and products are separated by an arrow showing which direction the reaction proceeds. Although chemical equations typically depict reactants forming products, in reality, chemical reactions can also proceed in the opposite direction. During a chemical reaction, no new atoms are created or destroyed (conservation of mass), but chemical bonds may be broken and formed between different atoms.

Chemical equations can be either unbalanced or balanced. An unbalanced chemical equation does not account for the conservation of mass, but it is often a good starting point because it lists the products and reactants, as well as the direction of the chemical reaction. Consider rust formation as an example. Rust is formed when iron reacts with oxygen in the air to create a new compound called iron oxide (rust). This chemical reaction can be expressed using the following unbalanced chemical equation, which can be written using either words or chemical symbols:

$$Fe + O \longrightarrow FeO$$

A more accurate description of a chemical reaction is given by writing a balanced chemical equation. A balanced chemical equation is written so that the number of atoms of each type of element is the same for both products and reactants. Coefficients before chemical species indicate quantities of reactants, while subscripts within a compound indicate the number of atoms of each element. Balanced chemical equations typically list the state of matter of each reactant (s for solid, l for liquid, g for gas). So, the balanced equation for the chemical reaction of rust formation becomes:

$$2\ Fe(s) + O_2(g) \longrightarrow 2\ FeO(s)$$

Examples of chemical reactions. There are millions of chemical reactions! Here are some examples: fire (combustion); baking a cake; cooking an egg; mixing baking soda and vinegar to produce salt and carbon dioxide gas.

1.7 Types of Chemical Reactions

Four main types of chemical reactions. Chemical reactions are everywhere. There are various methods for categorizing chemical reactions. Therefore, you may be prompted to identify the four, five, or six primary types of chemical reactions. As a chemist or chemical engineer, you may need to know the details of a very specific type of chemical reaction, but most reactions can be grouped into just a few categories. Determining the number of categories for chemical reactions poses a challenge. Generally, chemical reactions fall into either four, five, or six main types of reactions. Below is the standard classification.

Synthesis reaction (also known as a direct combination reaction). In this reaction, reactants combine to form a more complex product. Often there are two or more reactants with only a single product. The general reaction takes the form:

$$A + B \longrightarrow AB$$

The combination of iron and sulfur to form iron(II) sulfide is an example of a synthesis reaction:

$$Fe + S \longrightarrow FeS$$

Decomposition reaction (sometimes called an analysis reaction). In this kind of reaction, a molecule breaks into two or more smaller pieces. It's common to have one reactant and multiple products. The general chemical reaction is:

$$AB \longrightarrow A + B$$

The electrolysis of water into oxygen and hydrogen gas is an example of a decomposition reaction:

$$2 H_2O \longrightarrow 2 H_2 + O_2$$

Single displacement reaction (also called a single replacement reaction or substitution reaction). In this type of chemical reaction, one reactant ion changes place with another. The general form of the reaction is:

$$A + BC \longrightarrow B + AC$$

An example of a substitution reaction occurs when zinc combines with hydrochloric acid. The zinc replaces the hydrogen:

$$Zn + 2 HCl \longrightarrow ZnCl_2 + H_2$$

Double displacement reaction (also called a double replacement reaction or metathesis reaction). In this type of reaction, both cations and anions exchange places, according to the general reaction:

$$AB + CD \longrightarrow AD + BC$$

An example of a double displacement reaction occurs between sodium chloride and silver nitrate to form sodium nitrate and silver chloride:

$$NaCl(aq) + AgNO_3(aq) \longrightarrow NaNO_3(aq) + AgCl(s) \downarrow$$

Five main types of chemical reactions. You simply add one more category: **the combustion reaction**. The alternate names listed above still apply: (1) synthesis reaction; (2) decomposition reaction; (3) single displacement reaction; (4) double displacement reaction; (5) combustion reaction[5]. A general form of a combustion reaction is:

$$\text{hydrocarbon} + \text{oxygen} \longrightarrow \text{carbon dioxide} + \text{water}$$

Six main types of chemical reactions. The sixth type of chemical reaction is an **acid-base reaction**. The alternate names listed above still apply: (1) synthesis reaction; (2) decomposition reaction; (3) single displacement reaction; (4) double displacement reaction; (5) combustion reaction; and (6) acid-base reaction.

Other main categories of chemical reactions include oxidation-reduction (redox) reactions, isomerization reactions, and hydrolysis reactions. A chemical reaction is a process generally described by a chemical change in which the initiating substances (reactants) are different from the products. Chemical reactions tend to involve the motion of electrons, leading to the formation and breaking of chemical bonds. There are numerous types of chemical reactions, and they can be categorized in multiple ways. Below are some additional common reaction types apart from the ones previously mentioned.

Oxidation-reduction or redox reaction. In a redox reaction, the oxidation numbers of atoms are changed. Redox reactions may involve the transfer of electrons between chemical species. The reduction of I_2 to I^- and the oxidation of $S_2O_3^{2-}$ (thiosulfate anion) to $S_4O_6^{2-}$ (tetrathionate anion) provide an example of a redox reaction:

$$2\ S_2O_3^{2-}(aq) + I_2(aq) \longrightarrow S_4O_6^{2-}(aq) + 2\ I^-(aq)$$

Acid-base reaction. An acid-base reaction is a type of double displacement reaction that occurs between an acid and a base. The H^+ ion in the acid reacts with the OH^- ion in the base to form water and an ionic salt:

$$HA + BOH \longrightarrow H_2O + BA$$

The reaction between hydrobromic acid (HBr) and sodium hydroxide is an example of an acid-base reaction:

$$HBr + NaOH \longrightarrow NaBr + H_2O$$

Combustion. A combustion reaction is a type of redox reaction in which a combustible material combines with an oxidizer to form oxidized products and generate heat (exothermic reaction). Usually, in a combustion reaction oxygen combines with another compound to form carbon dioxide and water. An example of a combustion reaction is the burning of naphthalene:

$$C_{10}H_8 + 12\ O_2 \longrightarrow 10\ CO_2 + 4\ H_2O$$

Isomerization. In an isomerization reaction, the structural arrangement of a compound is changed but its net atomic composition remains the same.

Hydrolysis reaction. A hydrolysis reaction involves water. The general form for a hydrolysis reaction is:

$$X^-(aq) + H_2O(l) \rightleftharpoons HX(aq) + OH^-(aq)$$

Can a reaction be more than one type?

As you start adding more and more types of chemical reactions, you'll notice a reaction may fit into multiple categories. For example, a reaction may be both an acid-base reaction and a double displacement reaction.

1.8 The Study of Chemistry

Chemistry is a natural science that requires an objective viewpoint for proper understanding. It is the result of chemical scientific research and practice by countless chemists from ancient times to modern times. It contains some basic chemical concepts, basic theories, knowledge of elementary compounds, basic types of chemical reactions, classification of inorganic substances and their interrelationships. It is full of the principles and contents of materialistic dialectics. Chemistry plays an important role in industrial and agricultural production, national security and the modernization of science and technology. In fact, chemistry is integral to our everyday lives, from the clothes we wear to the food we eat, and even to our housing and transportation. Additionally, chemistry is a highly experimental field of science.

As a future practitioner in the chemical industry, you must not only study chemical theory and experimental skills, but also gain competence in performing chemical experiments, laying the groundwork for future production practices. Acquiring basic chemical laboratory skills is crucial for chemists, who must be able to operate chemical experiments effectively. Chemistry majors can gain a deeper understanding of dialectical materialism through the study of chemistry, while also establishing accurate views of life, values, and moral standards by exploring the patriotic sentiments of both domestic and foreign chemists presented in the history of chemistry. Chemistry students can benefit from this education by fostering curiosity and innovative attitudes toward scientific exploration. Additionally, this education can establish their observation abilities and critical thinking skills. Furthermore, they can develop their experimental and self-learning skills, providing a solid foundation for their future service in the chemical industry.

1.9 Scientific Method

All sciences, including the social sciences, use variations of what is called the scientific method, a systematic approach to research, as shown in Figure 1.1. Scientific methodology refers to the methodology of the natural sciences, which is the study of the general methods of the natural sciences, such as observation, experimentation, mathematical methods, and so on. For example, a psychologist who wants to know how noise affects people's ability to learn chemistry, and a chemist who wants to measure the heat given off when hydrogen gas burns in air, would follow roughly the same procedure in conducting their investigations. The first step is to carefully define the problem. The next step is to conduct experiments, make careful observations, and record information or data about the system — the part of the universe that is being studied.

The data obtained in a research study can be both qualitative, consisting of general observations about the system, and quantitative, consisting of numbers obtained from various measurements of the system[6]. Chemists generally use standardized symbols and equations to record their measurements and observations. This not only simplifies record keeping, but also provides a common basis for communicating with other chemists.

Once the experiments have been completed and the data recorded, the next step in the scientific method is interpretation, in which the scientist attempts to explain the observed phenomenon. Based on the data collected, the researcher formulates a **hypothesis**, a tentative explanation for a set of observations. Further experiments are designed to test the validity of the hypothesis in as many ways as possible, and the process begins again.

After a large amount of data has been collected, it is often desirable to summarize the information in a concise way, as a **law**. In science, a law is a succinct verbal or mathematical statement of a relationship between phenomena that is always the same under the same conditions[7]. For example, Sir Isaac Newton's second law of motion, which you may remember from high school science class, states that force equals mass times acceleration ($F = ma$). What this law means is that increasing the mass or acceleration of an object will always proportionally increase its force, and decreasing the mass or acceleration will always decrease the force.

Hypotheses that survive many experimental tests of their validity may evolve into theories. A **theory** is a unifying principle that explains a set of facts and/or the laws that underlie them. Theories are also constantly being tested. If a theory is disproved by experiment, it must be discarded or modified to be consistent with experimental observations. Proving or disproving a theory can take years, even centuries, in part because the necessary technology may not be available. The atomic theory we studied in college is a case in point. It took more than 2000 years to work out this fundamental principle of chemistry proposed by Democritus, an ancient Greek philosopher. A more recent example is the Big Bang theory of the origin of the universe.

Scientific progress rarely, if ever, occurs in a rigid, step-by-step fashion. Sometimes a law precedes a theory; sometimes it is the other way around. Two scientists may begin a project with exactly the same goal, but end up taking drastically different approaches. Scientists are human, after all, and the way they think and work is very much influenced by their background, training, and personality.

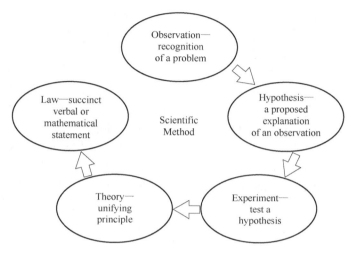

Figure 1.1　Schematic diagram of scientific methods in chemistry

The development of science is irregular and sometimes illogical. Great discoveries are usually the result of the cumulative contributions and experience of many workers, although the credit for formulating a theory or law is usually given to a single individual. There is, of course, an element of luck involved in scientific discovery, but it has been said that "chance favors the prepared mind"[8]. It takes an alert and well-trained person to recognize the significance of a serendipitous discovery and to take full advantage of it. Most of the time, the public hears only about spectacular scientific breakthroughs. For every success story, however, there are hundreds of cases in which scientists have spent years on projects that ultimately led to a dead end. And where positive results came only after many false starts, and at such a slow pace that they went unheralded. But even the dead ends contribute something to the ever-growing body of knowledge about the physical universe. It is the love of the search that keeps many scientists in the laboratory.

New Words and Expressions

overlap	[ˌəuvəˈlæp]	v. 重叠，交叠，（物体或时间上）部分重叠，（范围方面）部分重叠
branch	[brɑːntʃ]	n. 树枝，分支，分部，支流；v. 分开，分岔
characteristic	[ˌkærəktəˈrɪstɪk]	n. 特征，特点；adj. 独特的，特有的
caffeine	[ˈkæfiːn]	n. 咖啡因，咖啡碱
cube	[kjuːb]	n. 立方体，立方，立方形，三次幂，立方形的东西

英文	音标	释义
precursor	[prɪˈkɜːsə(r)]	n. 前身，先驱，先锋
alchemist	[ˈælkəmɪst]	n. 炼金术士
elixir	[ɪˈlɪksə(r)]	n. 长生不老药，灵丹妙药，万能药，圣水
scroll	[skrəʊl]	n.（供书写的）长卷纸，卷轴，纸卷；v. 滚屏，滚动
distillation	[ˌdɪstɪˈleɪʃn]	n. 蒸馏，蒸馏法
crystallization	[ˌkrɪstəlaɪˈzeɪʃ(ə)n]	n. 结晶化，具体化
sublimation	[ˌsʌblɪˈmeɪʃ(ə)n]	n. 升华，升华作用，高尚化，升华物
evaporation	[ɪˌvæpəˈreɪʃ(ə)n]	n. 蒸发
essential	[ɪˈsenʃ(ə)l]	adj. 必不可少的，基本的，精髓的；n. 必需品，要素，本质，基本知识
primitive	[ˈprɪmətɪv]	adj. 原始的，远古的；n. 词根
convenience	[kənˈviːniəns]	n. 方便，便利
telescope	[ˈtelɪskəʊp]	n. 望远镜，射电望远镜，无线电望远镜
paradigmatic	[ˌpærədɪɡˈmætɪk]	adj. 纵聚合关系的，典范的
sophisticated	[səˈfɪstɪkeɪtɪd]	adj. 见多识广的，复杂巧妙的，先进的，精密的，水平高的
carbon peak		碳达峰
carbon neutrality		碳中和
dual carbon		双碳
delve	[delv]	v. 钻研，探究，挖；n. 穴，洞
cutting-edge	[ˌkʌtɪŋ ˈedʒ]	adj. 领先的
individual	[ˌɪndɪˈvɪdʒuəl]	adj. 单独的，个别的，独特的，与众不同的；n. 个人，个体
drugs	[drʌɡz]	n. 毒品，药物
elucidate	[ɪˈluːsɪdeɪt]	v. 阐明，解释
macromolecule	[ˈmækrəʊmɒləkjuːl]	n. 大分子，高分子
pharmaceutical	[ˌfɑːməˈsuːtɪk(ə)l]	adj. 制药的，n. 药物
photocatalytic	[fəʊtəʊkætəˈlɪtɪk]	adj. 光催化（作用）的
prioritize	[praɪˈɒrətaɪz]	v. 优先处理
surpass	[səˈpɑːs]	v. 超过
supercapacitor	[suːpəkəˈpæsɪtər]	n. 超级电容器
adsorption	[ədˈzɔːpʃ(ə)n]	n. 吸附（作用）
indicator	[ˈɪndɪkeɪtə(r)]	n. 标志，迹象，指示器，显示器，指示剂
self-assembly	[ˌself əˈsembli]	n. 自组装；adj. 自行组装的
nanostructure	[ˈnænəʊstrʌktʃə(r)]	n. 纳米结构
biomedical	[ˌbaɪəʊˈmedɪk(ə)l]	adj. 生物医学的
industry	[ˈɪndəstri]	n. 行业，工业，生产制造，勤奋
agriculture	[ˈæɡrɪkʌltʃə(r)]	n. 农业，农学，农艺
polymer	[ˈpɒlɪmə(r)]	n. 聚合物，高分子
spectrometry	[spekˈtrɒmɪtri]	n. 光谱测定法
spectroscopy	[spekˈtrɒskəpi]	n. 光谱学

chromatography	[ˌkrəʊməˈtɒɡrəfi]	n.	色谱分析法
organism	[ˈɔːɡənɪzəm]	n.	生物，有机体
carbohydrate	[ˌkɑːbəʊˈhaɪdreɪt]	n.	糖类，碳水化合物
neurotransmitter	[ˈnjʊərəʊtrænzmɪtə]	n.	神经传导物质，神经递质
astrochemistry	[ˌæstrəʊˈkemɪstri]	n.	天体化学，太空化学
eco-friendly	[ˌiːkəʊˈfrendli]	adj.	对生态环境友好的，不妨害生态环境的
radiochemistry	[ˌreɪdiəʊˈkemɪstri]	n.	放射化学
radioisotope	[ˌreɪdiəʊˈaɪsətəʊp]	n.	放射性同位素
radiation	[ˌreɪdiˈeɪʃn]	n.	辐射，放射线，辐射的热（或能量等），放射疗法
programming	[ˈprəʊɡræmɪŋ]	n.	设计，编程，程序编制
rearrangement	[ˌriːəˈreɪndʒmənt]	n.	重新整理，重新排列
alter	[ˈɔːltə(r)]	v.	改变，改动，修改
naked	[ˈneɪkɪd]	adj.	裸露的（naked eye 裸眼，肉眼）
yield	[jiːld]	n.	产量，收益；v. 生产，投降，屈服
coefficient	[ˌkəʊɪˈfɪʃ(ə)nt]	n.	系数
vinegar	[ˈvɪnɪɡə(r)]	n.	醋，尖酸，刻薄
isomerization	[aɪsɒmaraɪˈzeɪʃn]	n.	异构化，异构化作用，异构化反应
hydrolysis	[haɪˈdrɒlɪsɪs]	n.	水解作用，水解
characterize	[ˈkærəktəraɪz]	v.	描述，刻画，成为……的特征，是……的典型
hydrochloric	[ˈhaɪdrəˈklɒrɪk]	adj.	氯化氢的，盐酸的
combustion	[kəmˈbʌstʃən]	n.	燃烧，氧化
acid-base	[ˈæsɪd beɪs]	n.	酸碱；adj. 酸碱的
naphthalene	[ˈnæfθəliːn]	n.	萘，卫生球，樟脑丸
dialectics	[ˌdaɪəˈlektɪks]	n.	辩证法（adj. 辩证的 dialectical）
materialism	[məˈtɪəriəlɪzəm]	n.	实利主义，物质主义，唯物论，唯物主义
patriotic	[ˌpeɪtriˈɑːtɪk]	adj.	爱国的
sentiment	[ˈsentɪmənt]	n.	情绪，观点，看法
domestic	[dəˈmestɪk]	adj.	国内的，家庭的，本国的
innovative	[ˈɪnəveɪtɪv]	adj.	创新的，革新的，采用新方法的，引进新思想的
tentative	[ˈtentətɪv]	adj.	试探性的，暂定的；n. 不确定的事物
hypothesis	[haɪˈpɒθəsɪs]	n.	假定，臆测（复数 hypotheses）
formulate	[ˈfɔːmjuleɪt]	v.	用公式表示，系统地阐述，制定
succinct	[səkˈsɪŋkt]	adj.	简明的，言简意赅的

Notes

1. The rapid development of increasingly sophisticated technology throughout the twentieth century has given us even greater means of studying things that cannot be seen with the naked eye.

整个20世纪，日益复杂的技术迅速发展，给了我们更好的方法来研究肉眼无法看到的事物。

2. As China entered the 21st century, it is highly likely that the chemical industry will play a significant role in science and technology to achieve the "dual carbon" goal, the assertion based on the Chinese government's continued commitment to achieve "carbon peak" and "carbon neutrality" in an effort to combat global climate change.
随着中国进入21世纪，化工行业极有可能在实现"双碳"目标的科技领域发挥重要作用，该"双碳"目标是中国政府为应对全球气候变化持续致力于实现"碳达峰"和"碳中和"而提出的。

3. Inorganic Chemistry Journals and Publications. There are numerous publications devoted to advances in inorganic chemistry. Journals include Inorganic Chemistry, Polyhedron, Journal of Inorganic Biochemistry, Dalton Transactions, and Bulletin of the Chemical Society of Japan.
无机化学期刊和出版物。有许多出版物致力于无机化学的进展。主要期刊有《无机化学》《多面体》《无机生物化学杂志》《道尔顿汇刊》《日本化学会公报》等。

4. In a chemical reaction, the atomic nucleus remains unchanged, while electrons can be shared or transferred to create or break down chemical bonds.
在化学反应中，原子核是不受影响的，但电子可以被转移或共享，从而断裂并形成化学键。

5. (1) synthesis reaction; (2) decomposition reaction; (3) single displacement reaction; (4) double displacement reaction; (5) combustion reaction.
（1）合成反应；（2）分解反应；（3）单置换反应；（4）双置换反应；（5）燃烧反应。

6. The data obtained in a research study can be both qualitative, consisting of general observations about the system, and quantitative, consisting of numbers obtained from various measurements of the system.
在研究中获得的数据可以是定性的，包括对系统的一般观察，也可以是定量的，包括从系统的各种测量中获得的数字。

7. In science, a law is a succinct verbal or mathematical statement of a relationship between phenomena that is always the same under the same conditions.
在科学中，定律是对在相同条件下总是相同的现象之间的关系的简洁的口头或数学陈述。

8. There is, of course, an element of luck involved in scientific discovery, but it has been said that "chance favors the prepared mind".
当然，在科学发现中也有运气的成分，但人们常说"机遇偏爱有准备的头脑"。

Trivial — Last but not the Least

How to read chemical formulas

作为化工科技人员在工作中必然会遇到化学分子式和化学方程式。认识这些分子式和方程式对于广大化学科技工作者并不困难，但是在一些特定场合仅仅认识是不够的，还必须会读。本补充知识点就是为了解决"读"方程式而设计的。下面介绍在会上做报告或者宣读科技文献的时候，怎样读化学分子式和化学方程式。其中，同一分子式可能有多种读法，本补充内容仅就一般情况加以介绍。

1. 凡是英语字母,不论大小写都读英语字母名称的读音,如 A/a、B/b 和 C/c 等分别读[eɪ]、[biː]和[siː]等;

2. 凡是数字一般都读作 one、two、three、four 等,如 K_2SO_4 读作:K-two-S-O-four;但是紧跟在圆括号后下方的数字要读作 twice、three times、four times 等;

3. 关于符号的读法

符号	读法	注释与案例
+	plus	[plʌs]
−	minus	[ˈmaɪnəs]
( )	round brackets/brackets/parentheses	[ˈbrækɪt]　[pəˈrenθəsiːz]　如$(OH)_2$的读法有三种,分别是 (1) (pause)—O—H—(pause)—twice; (2) open bracket—O—H—close bracket-twice;以及 (3) O—H—in brackets twice. (pause = 稍加停顿)
(	open bracket	
)	close bracket	
[ ]	square brackets	如是配合物,如$[Fe(CN)_6]K_4$则直接读 complexion 而不必读方括号
=	equals / is equal to	
→	give(s) / yield(s) / produce(s) / form(s) / become(s)	A ⟶ B + C 读作 A gives B plus C A + B ⟶ C + D 读作 A plus B gives C plus D / A and B give C and D / the reaction between A and B produces C and D
↔ / ⇌	reacts reversibly	
↑	evolved as a gas / a gas is evolved / gives off (a gas)	
↓	X is precipitated / X precipitates / gives an X precipitate	

4. 在英美出版的化学书中,在方程式中出现的分子式之后,往往加括号写明物质聚集态,如 s (solid)、g (gas)、l (liquid)或者 aq (aqueous);

5. 结晶水合物中的点一般读作 dot,如 $Na_2CO_3·10H_2O$ 读作 N—A—two—C—O—three—dot—ten—H—two—O;当然,这个 dot 也可以不读,用稍加停顿(pause)来代替(N—A—two—C—O—three—pause—ten—H—two—O);

6. 关于化学反应方程式中的催化剂的读法,如在加热条件下以 MnO_2 为催化剂发生反应,有 3 种读法:(1)in the presence of manganese dioxide as a catalyst on heating; (2) with a manganese dioxide catalyst under the influence of heat; (3) M—N—O—two-catalyst-with heat;

7. 以上所述均为一般性读法,如果当老师口授听写时,则必须表达得更加清楚,大小写也应该读清楚;例如:通常情况下,$Ca_3(PO_4)_2$ 可读成 C—A—three—pause—P—O—four-pause-twice;但当听写时,就应当读作:Capital C—lowercase A—three—open bracket—capital P—capital O—four—close bracket—twice;以上这些原则只有在多练习的基础上才能熟练掌握、灵活应用。

Unit 2
Working in the Lab

2.1 Safety in the Lab

The laboratory is inherently risky with the potential for fire hazards, hazardous chemicals, and dangerous procedures. It is important to follow laboratory safety guidelines to prevent accidents. The most basic laboratory safety rules are shown in Figure 2.1. The first safety guideline is to be aware of the potential hazards and to refrain from doing anything that you or your supervisor or trainer believe to be dangerous. If you think an activity is dangerous, discuss it first and do not proceed until appropriate precautions have been taken. Maintaining a habitable planet requires that we minimize the production of waste and dispose of it responsibly. Chemical recycling is practiced in industry for both economic and ethical reasons; it should be an important part of pollution control in your laboratory.

Familiarize yourself with laboratory safety precautions before working. Always wear safety glasses or goggles with side shields to protect your eyes from liquids and broken glass. Do not wear contact lenses as fumes may cause eye irritation. Protect your skin from spills and flames by wearing a lab coat made of flame-resistant material. Wear rubber gloves when handling concentrated acids. Never eat or drink in the laboratory. Handle organic solvents, concentrated acids, and concentrated ammonia in a fume hood to avoid exposure. This system prevents the escape of fumes into the laboratory by creating negative pressure through the intake of air. The incoming air dilutes the fumes before they are exhausted through the roof, ensuring safety and optimal air quality. It is highly recommended that you avoid generating large quantities of toxic fumes that could escape from the hood. Also, wear a mask when working with fine powders that could create a cloud of dust that could be inhaled.

Clean up spills immediately to prevent accidental exposure. For skin spills, rinse immediately with water. Know the location and use of emergency showers and eyewash stations for body or eye splashes. If the sink is closer to you than the eyewash station, use the sink first for eye splashes. Learn how to use fire extinguishers and emergency blankets to extinguish clothing fires, and have a fully stocked first aid kit on hand. Also, make sure you know how and where to get medical help in an emergency. Label all containers to identify their contents. An unlabeled bottle forgotten in a refrigerator or closet can lead to costly disposal problems because its contents must be analyzed before proper disposal procedures can be followed.

2.2 Disposal of Chemical Waste

Many of the chemicals we use are harmful to plants, animals, and humans if discarded carelessly. For each experiment, your supervisor should establish procedures for waste disposal. Options include: (1) pouring solutions down the drain and diluting them with tap water, (2) saving the waste for disposal in an approved landfill, (3) treating the waste to reduce the hazard and then pouring it down the drain or saving it for disposal in a landfill, and (4) recycling. Chemically incompatible wastes should never be mixed, and each waste container must be labeled to indicate the quantity and identity of its contents. Waste containers must indicate whether the contents are flammable, toxic, corrosive, reactive, or otherwise hazardous.

Figure 2.1   Working in the Lab

Some examples illustrate different approaches to laboratory waste management. Dichromate ($Cr_2O_7^{2-}$) is reduced to Cr^{3+} with sodium hydrogen sulfite ($NaHSO_3$), treated with hydroxide to produce insoluble $Cr(OH)_3$ and evaporated to dryness for disposal in a landfill. Waste acid is mixed with waste base to near neutral (as determined by pH paper) and then discharged down the drain. Waste iodate (IO_3^-) is reduced to I^- with $NaHSO_3$, neutralized with base and discharged. Waste Pb^{2+} solution is treated with sodium metasilicate (Na_2SiO_3) solution to precipitate insoluble $PbSiO_3$, which can be packaged for landfill. Waste silver or gold is treated to recover the metals. Toxic gases used in a fume hood are bubbled through a chemical trap or burned to prevent escape from the hood.

2.3   How to Deal with Bad Lab Partners?

Have you ever taken a lab course and had lab partners who didn't do their share of the work, broke equipment, or wouldn't work with you? This situation can be really hard, but there are steps you can take to make things better.

Talk to your lab partners. This may be harder than it sounds if your problem is that you and your lab partners don't speak the same language (which is relatively common in science and engineering), but you can improve your working relationship with your lab partners if you can explain to them what's bothering you. You also need to explain what you would like them to do and why you think these would make things better. Be prepared to compromise, as your lab partner may also want you to make some changes. Remember that you and your partner may come from very different cultures, even if you're from the same country. Avoid sarcasm or being "too nice" as there is a good chance you won't get your message across. If language is a problem, find an interpreter or draw pictures if necessary. If one or both of you don't want to be there, the work still needs to be done. If you know your partner won't do it, but your grade or career is on the line, you have to accept that you will do all the work. Now, you can still make it obvious that your partner is slacking off. On the other hand, if you both resent doing the work, it's reasonable to work out an arrangement. You may find that you work better together once you acknowledge that you hate the task.

If you have a lab partner who is willing to help but is incompetent or clumsy, try to find harmless tasks that allow the partner to participate without harming your data or your health. Ask for input, let the partner record data, and try to avoid stepping on toes. If the unsuspecting partner is a permanent fixture in your life, it's in your best interest to train them. Start with simple tasks, clearly explaining the steps, the reasons for certain actions, and the desired results. Be kind and helpful, but not condescending. If you're successful in your task, you'll gain a valuable ally in the lab, and maybe even a friend.

Maybe you and your lab partner had a fight, or there's a history. Maybe you just don't like each other. Unfortunately, it's not always possible to get out of a situation like this. You can ask your supervisor to reassign one or both of you, but you'll risk getting a reputation for being difficult to work with. If you decide to ask for a change, it's probably better to give a different reason for the request. If you absolutely must work together, try to set boundaries that limit how much you actually have to interact. Make your expectations clear so that you can both get the work done and get out.

It's better to try to work out problems with your lab partners than to ask your supervisor to intervene. However, there may be times when you need help or advice from someone higher up. This might be the case if you find that you can't meet a deadline or complete an assignment without more time or a change in the work dynamic. If you decide to talk to someone about your problems, present the situation calmly and without judgment. You have a problem; you need help finding a solution. This can be difficult, but it's a valuable skill to master. Difficulties with lab partners come with the territory. The social skills you can master in dealing with lab partners will help you whether you're taking a lab course or making a career out of lab work. No matter what you do, you'll need to learn to work well with others, including people who are incompetent, lazy, or simply don't want to work with you. If you want a career in science, you'll need to recognize and accept that you'll be a member of a team.

2.4 The Most Common Injuries in a Chemistry Lab

There are many hazards in a chemistry lab, such as chemicals, fragile objects, and open flames, so accidents are bound to happen. But an accident doesn't have to result in injury. Most common injuries can be prevented by minimizing accidents by being careful, wearing the right safety equipment and knowing what to do in an emergency. However, most of the time when people get hurt, it's either something they don't admit to or it's not a life-threatening event. What are your biggest risks? Here's an informal look at common injuries.

Eye injuries

Your eyes are at risk in the chemistry lab. If you normally wear contact lenses, wear glasses to reduce chemical exposure. Everyone should wear safety glasses. They protect your eyes from chemical splashes and accidental glass splinters. People get eye injuries all the time, either because they are lax about wearing safety glasses, the agent that causes the injury gets around the edge of the glasses, or they don't know how to use the eyewash properly. While cuts are more common in the laboratory, eye injuries are probably the most common serious injury.

Cuts from glassware

You can cut yourself if you are stupid enough to try to force a glass tube through a stopper with the palm of your hand. You can cut yourself breaking glassware or trying to clean up a mess. You can cut yourself on the sharp edge of a piece of broken glass. The best way to prevent this injury is to wear gloves, but it is still the most common injury, mainly because few people wear gloves all the time. Also, you lose dexterity when you wear gloves, so you may be more clumsy than usual.

Chemical irritation or burns

It's not just the skin on your hands that is at risk from chemical exposure, although that is the most common site of injury. You can inhale corrosive or reactive vapors. If you're particularly stupid, you can ingest harmful chemicals by swallowing liquid from a pipette, or (more commonly) by not cleaning up well enough after the lab and contaminating your food with traces of chemicals on your hands or clothes. Safety glasses and gloves will protect your hands and face. A lab coat protects your clothing. Don't forget to wear closed-toed shoes, because spilling acid on your foot is not a pleasant experience. It does happen.

Burns from heat

You can burn yourself on a hot plate, by accidentally picking up a piece of hot glassware, or by standing too close to a burner. If you have long hair, don't forget to tie it back. I've seen people set their bangs on fire in a Bunsen burner, so don't lean over a flame, no matter how short your hair is.

Mild to moderate poisoning

Chemical toxicity is an overlooked accident because symptoms can disappear in minutes to days. However, some chemicals or their metabolites can persist in the body for years, potentially causing organ damage or cancer. Accidentally drinking a liquid is an obvious source of poisoning, but many volatile compounds are dangerous when inhaled. Some chemicals are absorbed through the skin, so keep an eye out for spills.

Tips to prevent lab accidents

A little preparation can prevent most accidents. Here are some tips to keep yourself and others safe:

◆ Know (and follow) the safety rules for working in the lab. For example, if a particular fridge is marked "No Food", don't put your lunch there.

◆ Use your safety equipment. Wear your lab coat and goggles. Tie back long hair.

◆ Know the meaning of safety signs in the laboratory.

◆ Label containers of chemicals, even if they only contain water or other non-toxic materials. It's best to put a real label on a container, as crayon marks can be wiped off during handling.

◆ Make sure that safety equipment is maintained. Know the schedule for flushing the eyewash line. Check the ventilation of chemical hoods. Keep first aid kits in stock.

◆ Ask yourself if you're safe in the laboratory.

◆ Report problems. Whether it's faulty equipment or a minor accident, you should always report a problem to your immediate supervisor. If no one knows there's a problem, it's unlikely to be fixed.

2.5　The Lab Notebook for Chemistry Lab Report

Lab reports are time-consuming for students, so why are they so important? There are two main reasons. First, a lab report is an orderly way to report the purpose, procedure, data, and results of an experiment. In essence, it follows the scientific method. Second, lab reports are easily adapted into papers for peer-reviewed publication. For students who are serious about a career in science, a lab report is a stepping stone to submitting work for review. Even if the results aren't published, the report is a record of how an experiment was conducted, which can be valuable for further research.

The critical functions of your lab notebook are to describe what you did and what you observed, and to be understandable to a stranger. The biggest mistake even experienced scientists make is writing incomplete or incomprehensible notebooks. Using complete sentences is an excellent way to avoid incomplete descriptions.

Beginners often find it useful to write a complete description of an experiment, with sections on purpose, methods, results, and conclusions. Setting up a notebook to record numerical data before coming to the lab is an excellent way to prepare for an experiment. It is good practice to write a balanced chemical equation for each reaction you use. This practice will help you understand what you are doing and may show you what you do not understand about what you are doing. The measure of scientific "truth" is the ability of different people to reproduce an experiment. A good lab notebook will record everything that was done and what was observed, and will allow you or anyone else to repeat the experiment. Record the names of computer files where programs and data are stored in your notebook. Keep hard copies of important data in your notebook. The life of a printed page is an order of magnitude (or more) greater than the life of a computer disk.

A lab notebook is the most important permanent record of your research and experiments. Note that if you are taking an AP lab course, you will need to submit an appropriate lab notebook in order to receive AP credit at most colleges and universities. Here is a list of **guidelines** that explain how to keep a lab notebook.

The notebook must be permanently bound. It should not be loose-leaf or 3-ring binder. Never tear a page out of the lab notebook. If you make a mistake, you can cross it out, but you should not remove any pages or parts of pages from your notebook. If you cross out an error, it should still be legible. You should explain the reason for the cross-out, and you should initial and date it. It is not acceptable to make notes in pencil or erasable ink.

Organization is the key to a good record book. Print your name, contact information, date, and any other pertinent information on the cover of your record book. Some record books require you to write some of this information on each page of the book. If your book is not prenumbered, number each page. Usually the numbers are in the top outside corner, and both the front and back of each page are numbered. Your teacher may have a rule about numbering. If so, follow their instructions. It's also a good idea to reserve the first few pages for a table of contents. To keep everything organized and simple, start a new page for each experiment.

Be accurate in your record keeping. This is a record of the lab work you have done during the semester or year, so it needs to be thorough. For each experiment, record the date(s) and list lab partners, if applicable. Record all the information in real time. Don't wait to record information. It may be tempting to record data elsewhere and then transcribe it into your lab notebook, usually because it makes the notebook look neater, but it's important to record it immediately. Include diagrams, photos, graphs, and similar information in your lab notebook. Typically, these are taped in or include a pocket for a data chip. If you need to keep some data in a separate book or other location, note the location in your lab notebook and cross-reference it with the appropriate lab notebook page numbers wherever the data is stored. Don't leave gaps or spaces in the record book. If you have a large blank space, cross it out. This is so that no one can go back and add incorrect information later.

Science lab report template

When writing a lab report, it can be helpful to have a template to work from. This science fair project lab report template allows you to fill in the blanks, making the writing process easier. Use the template with the instructions for writing a science lab report to ensure success.

Lab report headings

Generally, these are the headings you'll use in a lab report, in this order:
✧ Title
✧ Date
✧ Lab partners
✧ Purpose
✧ Introduction
✧ Materials
✧ Procedure
✧ Data
✧ Results
✧ Conclusion
✧ References

Overview of the parts of a lab report

Here's a quick look at the types of information you should include in each section of the lab report and a guide to how long each section should be. It's a good idea to look at other lab reports submitted by other groups that received a good grade or are well respected. In a classroom setting, it takes a long time to grade lab reports. You don't want to repeat an error if you can avoid it!

✧ **Title:** This should accurately describe the experiment. Don't try to be cute or funny.
✧ **Date:** This can be the date you did the experiment or the date you finished the report.
✧ **Lab partners:** Who helped you with the experiment? Give their full names. If they represent other schools or institutions, give them credit too.
✧ **Purpose:** This is sometimes called the objective. It is either a one-sentence summary of why the experiment or product was done, or a single paragraph.
✧ **Introduction:** Describe why the topic is of interest. The introduction may be a paragraph or a single page. The last sentence is usually a statement of the hypothesis tested.
✧ **Materials:** List the chemicals and special equipment used in this experiment. Ideally, this section should be detailed enough for another person to repeat the experiment.
✧ **Procedure:** Describe what you did. This can be a single paragraph or one or more pages.
✧ **Data:** List the data you obtained, before any calculations. Tables and graphs are good.
✧ **Results:** If you did any calculations on the data, these are your results. An error analysis is usually included here, although it may be a separate section.
✧ **Conclusion:** State whether the hypothesis was accepted or the project was a success. It's a good idea to suggest avenues for further study.
✧ **References:** Cite any resources or publications you have used. Did you consult a paper that was in any way related to the project? Give credit. References are needed for all facts except those that are readily available to the report's intended audience.

2.6 Basic Laboratory Apparatus and Manipulation

2.6.1 Apparatus

Analytical balance

An electronic balance uses an electromagnet to balance the load on the pan. A typical analytical balance, as shown in Figure 2.2, has a capacity of 100 - 200 g and a sensitivity of 0.01 - 0.1 mg. Sensitivity is the smallest change in mass that can be measured. The microbalance weighs milligram amounts with a sensitivity of 0.1 μg.

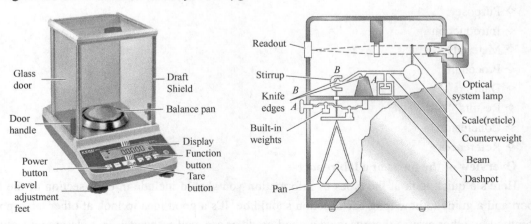

Figure 2.2 The electronic balance (left) and the single-pan mechanical analytical balance (right)

To weigh a chemical, first place a clean container on the weighing pan. The mass of the empty container is called the tare. On most balances, you can press a button to reset the tare to zero. Add the chemical to the container and read the new mass. If there is no automatic tare, subtract the tare mass from the mass of the filled container. Never place chemicals directly on the weighing pan. This precaution protects the balance from corrosion and allows you to recover all of the weighed chemical.

An alternative method called "weighing by difference" is used routinely by many people and is necessary for hygroscopic reagents that rapidly absorb moisture from the air. First, weigh a capped bottle of dry reagent. Then quickly pour some reagent from the weighing bottle into a recipient. Replace the cap on the weighing bottle and weigh again. The difference is the mass of reagent dispensed from the weighing bottle.

How does it work? The object on a balance pushes the pan down with a force equal to $m \cdot g$, where m is the mass of the object and g is the acceleration of gravity. The electronic balance uses the electromagnetic force to return the pan to its original position. The electric current required to generate the force is proportional to the mass, which is indicated on a digital display.

When a mass is placed on the pan, the zero detector senses a shift and sends an error signal to the circuit that generates a correction current. This current flows through the coil under the balance pan, creating a magnetic field that is repelled by a permanent magnet under the pan. As the displacement decreases, the output of the zero detector decreases. The correction current required to return the pan to its original position is proportional to the mass on the pan.

The older single pan mechanical balance uses standard masses and a balance beam suspended from a sharp knife edge to measure the mass of the object on the balance pan. The mass of the pan hanging from the balance point (another knife edge) on the left is balanced by a counterweight on the right. An object placed on the pan pushes the pan down. Knobs are then adjusted to remove weights from a beam above the pan and hidden inside the balance. The beam is restored to nearly its original position when the masses removed from the beam are nearly equal to the mass of the object on the pan. The slight difference from the original position is shown on an optical scale, whose reading is added to that of the knobs.

A mechanical balance should be in the locked position when loading or unloading the pan, and in the semi-locked position when selecting weights. This practice prevents sudden forces from wearing down the knife edges and reducing the sensitivity of the balance.

Buret (Burette)

The buret is a precision-manufactured glass tube with graduations that allow you to measure the volume of liquid delivered through the stopcock (valve) at the bottom. The 0 mL mark is near the top. If the initial liquid level is 0.85 mL and the final level is 27.98 mL, then you have dispensed 27.98 − 0.85 = 27.13 mL. Class A burets (the most accurate grade) are certified to meet the tolerances. If the reading on a 50 mL buret is 27.98 mL, the true volume can be anywhere from 28.03 to 27.93 mL and still be within the tolerance of ±0.05 mL.

To accurately locate the end of a titration, dispense less than one drop at a time from the buret near the end point (one drop from a 50 mL buret is approximately 0.05 mL). To dispense a fraction of a drop, gently open the stopcock until part of a drop hangs from the buret tip (some people prefer to rotate the stopcock quickly through the open position to dispense part of a drop). Then touch the inside glass wall of the receiving flask to the buret tip to transfer the drop to the wall of the flask. Carefully invert the flask so that the main body of the liquid washes over the newly added drop. Swirl the flask to mix the contents. Toward the end of a titration, invert and swirl the flask frequently to ensure that droplets on the wall containing unreacted analyte contact the bulk solution.

Liquid should flow evenly down the wall of a buret. The tendency of liquid to stick to the glass is reduced by draining the buret slowly (<20 mL/min). If many droplets adhere to the wall, clean the buret with detergent and a buret brush. If this cleaning is not sufficient, soak the buret in peroxydisulfate - sulfuric acid cleaning solution, which will "eat" clothing and people as well as grease in the buret. Never soak volumetric glassware in alkaline solutions, which will attack the glass. A 5% (by mass) NaOH solution at 95℃ will dissolve Pyrex glass at a rate of 9 μm/h.

Errors may be caused by failure to purge the air bubble often found just below the stopcock. If the bubble fills with liquid during the titration, some of the volume drained from the graduated portion of the buret has not reached the titration vessel. The bubble can be removed by draining the buret for a second or two with the stopcock wide open. A stubborn bubble can be dislodged by shaking the buret abruptly while draining it into a sink.

When refilling a buret with new solution, it is a good idea to rinse the buret several times with small portions of new solution, discarding each wash. It is not necessary to fill the buret with wash solution. Simply tilt the buret to expose all surfaces to the wash solution. The same technique should be used with any vessel, such as a spectrophotometer cuvette or pipette, that is reused without drying.

The digital titrator is convenient for use in the field where samples are collected. The counter indicates how much reagent has been dispensed. The 1% accuracy is 10 times worse than that of a glass buret, but many measurements do not require higher accuracy. The battery-operated electronic buret fits over a reagent bottle and dispenses up to 99.99 mL in 0.01 mL increments. For titrations requiring the highest precision, the mass of the reagent is dispensed from a buret or syringe instead of the volume. Mass can be measured more accurately than volume.

Volumetric flask

A volumetric flask is calibrated to contain a given volume of solution at 20℃ when the bottom of the meniscus is aligned with the center of the mark on the neck of the flask. Most flasks are marked "TC 20℃", which means to contain at 20℃. The temperature of the container is important because both liquid and glass expand when heated. To use a volumetric flask, dissolve the desired mass of reagent in the flask by swirling with less than the final volume of liquid. Then add more liquid and swirl again. When mixing two different liquids, there is usually a small change in volume. The total volume is not the sum of the two mixed volumes. By swirling the liquid in a nearly full flask before the liquid reaches the thin neck, you minimize the volume change when the last liquid is added. For good control, add the last drops of liquid with a pipette, not a squeeze bottle. Finally, hold the cap firmly in place and invert the flask 20 times to mix well.

Pipettes and syringes

Pipettes dispense predetermined volumes of liquid. The transfer pipette is calibrated to deliver a precise volume. The last drop should not be blown out or allowed to drip. The graduated pipette is calibrated like a buret. It is used to dispense a variable volume, such as 4.9 mL, by starting dispensing at the 1.0 mL mark and ending at the 5.9 mL mark. The transfer pipette is more accurate, with tighter tolerances.

How to use a transfer pipette? Use a rubber bulb or other pipette aspirator, not your mouth, to aspirate liquid past the calibration mark. Discard one or two pipette volumes of liquid to rinse traces of previous reagents from the pipette. After aspirating the third volume past the calibration mark, quickly replace the bulb with your index finger on the end of the pipette. Gently pressing the pipette against the bottom of the vessel while removing the rubber bulb will help prevent liquid from draining below the mark while you place your finger in place (alternatively, you may use an automatic aspirator that remains attached to the pipette)*. Wipe excess liquid from the outside of the pipette with a clean tissue. Touch the tip of the pipette to the side of a beaker and drain the liquid until the bottom of the meniscus just reaches the center of the mark. Touching the beaker will draw liquid from the pipette without leaving any part of a drop hanging when the liquid reaches the calibration mark. Transfer the pipette to a collection vessel and allow it to drain by gravity while holding the tip against the wall of the vessel. After the liquid stops flowing, hold the pipette against the wall for a few seconds to complete emptying. Do not blow out the last drop; the pipette should be nearly vertical at the end of dispensing. When you are finished with a pipette, rinse it with distilled water or soak it until you are ready to clean it. Solutions should never be allowed to dry inside a pipette as it is very difficult to remove internal deposits.

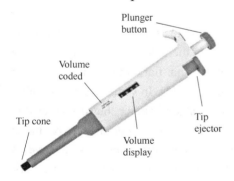

Micropipette

Micropipettes dispense volumes from 1 to 1000 μL (1 μL = 10^{-6} L). The liquid is contained in the disposable polypropylene tip, which is stable to most aqueous solutions and many organic solvents except chloroform ($CHCl_3$). The tip is not resistant to concentrated nitric acid or sulfuric acid.

When using a micropipette, place a fresh tip tightly on the tip cone. Store tips in their packaging or in the dispenser to avoid contaminating the tips with your fingers. Set the desired volume with the knob at the top of the pipette. Depress the piston to the first stop corresponding to the selected volume. Holding the pipette vertically, immerse the tip 3 - 5 mm into the reagent solution and slowly release the plunger to aspirate the liquid. Remove the tip from the liquid by sliding it along the wall of the vessel to remove liquid from the outside of the tip. To dispense liquid, touch the micropipette tip to the wall of the vessel and gently depress the plunger to the first stop. Wait a few seconds to allow liquid to run down the wall of the pipette tip, then depress the plunger further to dispense the last of the liquid. It is a good idea to clean and wet a fresh tip by taking up and discarding two or three squirts of reagent. The tip may be discarded or thoroughly rinsed with a squeeze bottle and reused. The amount of liquid transferred to the tip depends on the angle at which the pipette is held and how far below the surface of the liquid the tip is held during transfer. Everyone achieves slightly different levels of precision and accuracy with a micropipette.

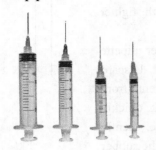

Syringe

Microliter syringes are available in sizes from 1 to 500 μL and have an accuracy and precision of close to 1%. The digital dispenser improves accuracy and precision to 0.5%. When using a syringe, take up and discard several volumes of liquid to wash the glass walls and remove air bubbles from the cylinder. The steel needle is corroded by strong acid and will contaminate strongly acidic solutions with iron.

Beaker

Beakers are the workhorse glassware of any chemistry lab. They come in a variety of sizes and are used to measure volumes of liquid. Beakers aren't very accurate. Some aren't even marked with volume measurements. A typical beaker is accurate to within about 10%. In other words, a 250 mL beaker will hold (250±25) mL of liquid. A 1 L beaker will be accurate to about 100 mL of liquid. The flat bottom of a beaker makes it easy to place on flat surfaces such as a lab bench or hot plate. The spout makes it easy to pour liquids into other containers.

Finally, the wide opening makes it easy to add materials to the beaker. For these reasons, beakers are often used for mixing and transferring liquids.

Erlenmeyer flask

There are several types of flasks. One of the most common in a chemistry lab is the Erlenmeyer flask. This type of flask has a narrow neck and flat bottom. It's good for stirring, storing, and heating liquids. In some situations, either a beaker or an Erlenmeyer is a good choice, but if you need to seal a container, it's much easier to put a stopper in an Erlenmeyer or cover it with parafilm than it is to cover a beaker. Erlenmeyer flasks come in several sizes. Like beakers, these flasks may or may not be marked with a volume. They are accurate to about 10%.

Test tube

Test tubes are good for collecting and holding small samples. They aren't typically used to measure precise volumes. Test tubes are relatively inexpensive compared to other types of glassware. Those designed to be heated directly with a flame are sometimes made of borosilicate glass, but others are made of less sturdy glass and sometimes plastic. Test tubes usually don't have volume markings. They are sold by size and may have either smooth

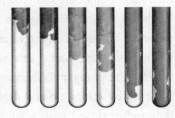

openings or lips.

Centrifuge and fume hood

A centrifuge is a motorized piece of laboratory equipment that spins liquid samples to separate their components. Centrifuges come in two main sizes, a benchtop version, often called a microcentrifuge, and a larger floor model.

A fume hood is a piece of laboratory equipment designed to limit exposure to hazardous fumes. The air inside the fume hood is either vented to the outside or filtered and recirculated.

Other apparatus

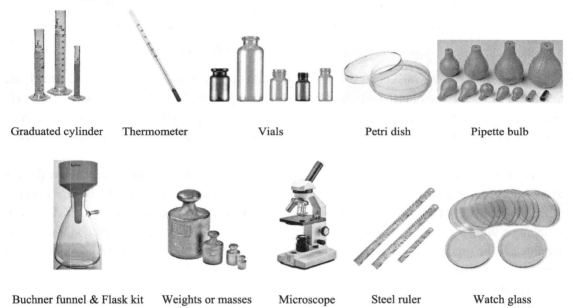

Graduated cylinder Thermometer Vials Petri dish Pipette bulb

Buchner funnel & Flask kit Weights or masses Microscope Steel ruler Watch glass

2.6.2 Manipulation

Filtration

In gravimetric analysis, the mass of product from a reaction is measured to determine how much unknown was present. Precipitates from gravimetric analyses are collected by filtration, washed, and then dried. Most precipitates are collected in a fritted glass funnel which is also called a Gooch filter crucible, with suction applied to speed filtration. The porous glass plate in the funnel allows liquid to pass through but retains solids. The empty funnel is first dried at 110°C and weighed. After the solid is collected and dried again, the funnel and its contents are weighed a second time to determine the mass of the solid collected. The liquid from which a substance precipitates or crystallizes is called the mother liquor. The liquid that passes through the filter is called the filtrate, as depicted in Figure 2.3.

In some gravimetric methods, ignition (heating at high temperature over a burner or in a furnace) is used to convert a precipitate to a known, constant composition. For example, Fe^{3+} precipitates as hydrous iron oxide, $FeOOH \cdot xH_2O$, with a variable composition. Ignition converts it to pure Fe_2O_3 prior to weighing. If a precipitate is to be ignited, it is collected on ashless filter paper, which leaves little residue when burned.

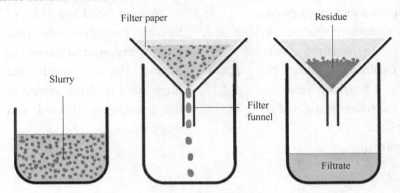

Figure 2.3　Schematic diagram of filtration

To use filter paper with a conical glass funnel, fold the paper into quarters, tear off one corner to allow a firm fit in the funnel, and place the paper in the funnel. The filter paper should fit snugly and be moistened with a little distilled water. When liquid is poured in, a continuous stream of liquid should fill the stem of the funnel. The weight of the liquid in the stem will help speed filtration.

For filtration, pour the slurry, which is a suspension of solid in liquid, of the precipitate down a glass rod to prevent splashing. Particles adhering to the beaker or rod can be dislodged with a rubber policeman, which is a flattened piece of rubber on the end of a glass rod. Use a jet of suitable washing liquid from a squeeze bottle to transfer particles from the rubber and glassware to the filter. If the precipitate is to be ignited, wipe the particles remaining in the beaker onto a small piece of moist filter paper. Place this paper in the filter to be ignited.

Drying

Reagents, precipitates, and glassware are conveniently dried in an oven at 110℃ (some chemicals require other temperatures.) Anything you put in the oven should be labeled. Use a beaker and a watch glass to minimize dust contamination during drying. It is good practice to cover all vessels on the bench top to prevent dust contamination.

The mass of a gravimetric precipitate is measured by weighing a dry, empty filter crucible before the procedure and reweighing the same crucible filled with dry product after the procedure. To weigh the empty crucible, first bring it to "constant mass" by drying it in an oven for 1 hour or more and then cooling it in a desiccator for 30 minutes. Weigh the crucible and reheat for about 30 minutes. Cool and reweigh. If the successive masses are within ±0.3 mg, the filter has reached "constant mass". You can use a microwave oven instead of an electric oven to dry reagents and crucibles. Try an initial heating time of 4 minutes, followed by 2 minutes of heating. Allow 15 minutes for cooling before weighing.

A **desiccator** is a closed chamber containing a drying agent called desiccant. The lid is greased to create an airtight seal, and the desiccant is placed in the bottom under the perforated plate. A useful desiccant is 98% sulfuric acid. After placing a hot object in the desiccator, leave the lid cracked open for one minute until the object has cooled slightly. This practice prevents the lid from popping open when the air inside heats up. To open a desiccator, slide the lid sideways rather than trying to pull it straight up.

New Words and Expressions

inherently	[ɪnˈherəntli]	adv. 固有的
refrain	[rɪˈfreɪn]	v. 避免，克制，抑制，节制
supervisor	[ˈsuːpəvaɪzə(r)]	n. 监督人，主管人，指导者
splash	[splæʃ]	v. 泼洒，朝……上泼（或溅）；n. 溅洒后留下的污渍，溅上的液体
precaution	[prɪˈkɔːʃn]	n. 预防，预防措施，防备
recycle	[ˌriːˈsaɪkl]	v. 回收利用，再利用，（将可回收物品）送入回收加工厂
compromise	[ˈkɒmprəmaɪz]	v.（为达成协议而）妥协，折中，让步；n. 妥协，折中，和解
sarcasm	[ˈsɑːkæzəm]	n. 讽刺，挖苦，嘲讽
interpreter	[ɪnˈtɜːprətə(r)]	n. 口译译员，解释程序，口译工作者
ethical	[ˈeθɪkl]	adj. 道德的，伦理的，合乎道德（或规矩）的
goggle	[ˈɡɒɡl]	n. 护目镜，风镜，游泳镜（常用复数形式 goggles）
lens	[lenz]	n. 透镜，镜片，（眼球的）晶状体
splinter	[ˈsplɪntə(r)]	n. 碎片，（木头、金属、玻璃等的）尖碎片，尖细条
lax	[læks]	adj. 松懈的，松弛的，不严格的，马虎的
dexterity	[dekˈsterəti]	n. 灵巧，灵活，（思维）敏捷，（手）熟练
flame-resistant	[fleɪmrɪˈzɪstənt]	adj. 阻燃的，耐火的
ingest	[ɪnˈdʒest]	v. 摄入，咽下
metabolite	[mɪˈtæbəˌlaɪt]	n. 代谢物，代谢产物
AP course		Advanced Placement Course 大学先修课程
rubber	[ˈrʌbə(r)]	n. 橡胶，橡皮，黑板擦
ammonia	[əˈməʊniə]	n. 氨，氨水
fume	[fjuːm]	v. 发怒，生气，冒气（或烟、汽），熏（尤指木材）
respirator	[ˈrespəreɪtə(r)]	n. 口罩，呼吸器，防毒面具
extinguisher	[ɪkˈstɪŋɡwɪʃə(r)]	n. 灭火器
dispose	[dɪˈspəʊz]	v. 处理，配置，使适应，使有倾向
accompany	[əˈkʌmpəni]	v. 伴随，陪同，陪伴，与……同时发生
dilute	[daɪˈluːt]	v. 稀释，降低，冲淡，削弱，使降低效果
flammable	[ˈflæməbl]	adj. 易燃的，可燃的，可燃性的；n. 易燃物
toxic	[ˈtɒksɪk]	adj. 有毒的，引起中毒的，中毒的；n. 有毒物质，毒药
corrosive	[kəˈrəʊsɪv]	adj. 腐蚀的，侵蚀性的；n. 腐蚀物
hazardous	[ˈhæzədəs]	adj. 危险的，有害的

illustrate	[ˈɪləstreɪt]	v. 加插图于，说明，阐明，证明，证实
dichromate	[daɪˈkrəʊmeɪt]	n. 重铬酸盐
hydrogen	[ˈhaɪdrədʒən]	n. 氢，氢气
hydroxide	[haɪˈdrɒksaɪd]	n. 氢氧化物，羟基化合物
insoluble	[ɪnˈsɒljəb(ə)l]	adj. 不能解决的，不能溶解的（soluble adj. 可溶的，可解决的）
evaporate	[ɪˈvæpəreɪt]	v. 蒸发，消失，失去水分
dryness	[ˈdraɪnəs]	n. 干燥，冷淡，干燥无味
discharge	[dɪsˈtʃɑːdʒ]	v. 排出，卸货
metasilicate	[ˌmetəˈsɪlɪˌkeɪt]	n. 硅酸盐
bubble	[ˈbʌb(ə)l]	v. 冒气泡，沸腾，充溢
slack	[slæk]	v.（使）松弛，放慢，熟化（石灰）（slack off 懈怠）
clumsy	[ˈklʌmzi]	adj. 笨拙的
toe	[təʊ]	n. 脚趾（step on one's toes 触怒某人，惹怒）
unsuspecting	[ˌʌnsəˈspektɪŋ]	adj. 不怀疑的，未猜想到……的
reassign	[ˌriːəˈsaɪn]	v. 再分配，再指定
minimize	[ˈmɪnɪmaɪz]	v. 将……减到最少
irritation	[ˌɪrɪˈteɪʃ(ə)n]	n. 恼怒，生气，（身体某部位的）疼痛，刺激（作用），兴奋
burner	[ˈbɜːnə(r)]	n. 火炉，燃烧炉，灯头（Bunsen burner 本生灯，科学实验用煤气灯）
volatile	[ˈvɒlətaɪl]	adj. 易变的，（液体或固体）易挥发的，易气化的；n. 挥发物
non-toxic	[nɒn ˈtɒksɪk]	adj. 无毒的；n. 无毒性
crayon	[ˈkreɪən]	n. 彩色粉笔（或炭笔、蜡笔），彩色粉笔（或炭笔、蜡笔）画
time-consuming	[ˈtaɪm kənsjuːmɪŋ]	adj. 耗时的，旷日持久的
peer-reviewed	[ˌpɪər rɪˈvjuːd]	adj. 经同行评议的
guideline	[ˈgaɪdlaɪn]	n. 指导方针
loose-leaf	[ˌluːs ˈliːf]	adj. 活页的
legible	[ˈledʒəb(ə)l]	adj.（字迹，印刷）清楚的，易读的
hygroscopic	[haɪgrəˈskɒpɪk]	adj. 吸湿的，湿度计的，易潮湿的
reagent	[rɪˈeɪdʒənt]	n. 试剂
moisture	[ˈmɔɪstʃə(r)]	n. 潮气，水分
circuit	[ˈsɜːkɪt]	n. 电路，回路；v.（绕……）环行
beam	[biːm]	n. 梁，横梁，光线，光柱；v. 定向发出（无线电信号），照耀
arrest	[əˈrest]	v. 逮捕，拘留，阻止，抑制，吸引，心跳停止
edge	[edʒ]	n. 边缘，边界，优势

burst	[bɜːst]	v.	爆裂，突然出现，爆满，猛冲
stopcock	[ˈstɒpkɒk]	n.	活塞，开关，止水栓，水龙头
tolerance	[ˈtɒlərəns]	n.	忍受，忍耐力，耐药力，耐受性，公差，容许偏差
parallax	[ˈpærəlæks]	n.	视差
meniscus	[məˈnɪskəs]	n.	半月板，弯月面，新月形物
thickness	[ˈθɪknəs]	n.	厚度，粗细，层，浓度，密度，最厚（或最深）处
negligible	[ˈneɡlɪdʒəb(ə)l]	adj.	微不足道的，不值一提的
dispense	[dɪˈspens]	v.	发放，分配，提供，施予，配（药），发（药）
flask	[flɑːsk]	n.	烧瓶，长颈瓶，细颈瓶，酒瓶，携带瓶
analyte	[ˈænəˌlaɪt]	n.	分析物，被分析物
tendency	[ˈtendənsi]	n.	经常性行为，偏好，趋势，趋向，倾向
adhere	[ədˈhɪə(r)]	v.	黏附，附着，遵守，遵循（规定或协议），拥护，持有（观点或信仰）
detergent	[dɪˈtɜːdʒənt]	n.	洗涤剂，去污剂；adj. 使清洁的，（与）清洁添加剂（有关）的
Pyrex	[ˈpaɪreks]	n.	派热克斯玻璃（一种耐热玻璃）
purge	[pɜːdʒ]	v.	清洗，清除，删除（不需要的东西）
pipette	[pɪˈpet]	n.	移液管，吸移管
digital	[ˈdɪdʒɪt(ə)l]	adj.	数字的，数码的，数字显示的
titrator	[ˈtaɪtreɪtə]	n.	滴定仪，滴定器
volumetric	[ˌvɒljuˈmetrɪk]	adj.	体积的，容积的，测定体积的
syringe	[sɪˈrɪndʒ]	n.	注射器，针筒，吸管，喷射器
aspirator	[ˈæspɪreɪtə(r)]	n.	抽吸器，吸引器，吸气器
micropipette	[ˌmaɪkrəʊpɪˈpet]	n.	微量吸液管
polypropylene	[ˌpɒliˈprəʊpəliːn]	n.	聚丙烯
chloroform	[ˈklɒrəfɔːm]	n.	氯仿，三氯甲烷
microliter	[ˌmaɪkrəʊˈliːtə(r)]	n.	微升
manipulation	[məˌnɪpjuˈleɪʃn]	n.	操纵，（对储存在计算机上的信息的）操作，处理
filtrate	[ˈfɪltreɪt]	v.	过滤，筛选；n. 滤液（filtration n. 过滤，筛选）
ignition	[ɪɡˈnɪʃ(ə)n]	n.	着火，点燃，点火装置（ignite v. 点燃，（使）燃烧，着火）
flatten	[ˈflætn]	v.	平整，打倒
crucible	[ˈkruːsɪb(ə)l]	n.	坩埚，严酷的考验
desiccator	[ˈdesɪkeɪtə(r)]	n.	干燥器，干燥剂
isotropic	[ˌaɪsəʊˈtrɒpɪk]	adj.	各向同性的，等方性的
anisotropic	[ænˌaɪsəʊˈtrɒpɪk]	adj.	各向异性的，非均质的
opacity	[əʊˈpæsəti]	n.	不透明，不传导，暧昧

grain	[greɪn]	n. 谷粒
elongate	[ˈiːlɒŋgeɪt]	v. 拉长，延长，变长
covalent	[ˌkəʊˈveɪlənt]	adj. 共价的，共有原子价的
flammability	[ˌflæməˈbɪlɪti]	n. 可燃性，易燃性
combustible	[kəmˈbʌstəbl]	adj. 易燃的，易激动的，燃烧性的
kerosene	[ˈkerəsiːn]	n. 煤油，火油
alcohol	[ˈælkəhɒl]	n. 含酒精饮品，酒，酒精，乙醇
enthalpy	[enˈθælpɪ]	n. 焓（热函），热含量
electronegativity	[ɪˌlektrəʊnegəˈtɪvɪti]	n. 电负性，电负度
acidity	[əˈsɪdəti]	n. 酸度，（言辞的）尖酸，酸性，胃酸过多
basicity	[beɪˈsɪsɪti]	n. 碱度
odorless	[ˈəʊdələs]	adj. 没有气味的
pungent	[ˈpʌndʒənt]	adj. 辛辣的，刺激性的，刺鼻的，苦痛的，尖刻的
penetrate	[ˈpenətreɪt]	v. 穿透，贯穿
solubility	[ˌsɒljuˈbɪləti]	n. 溶解度，可解决性
aqueous	[ˈeɪkwɪəs]	adj. 水的，水般的
ductile	[ˈdʌktaɪl]	adj. 柔软的，易教导的，易延展的
malleable	[ˈmælɪəb(ə)l]	adj. 有延展性的，易塑形的，容易改变的，易受影响的
conductor	[kənˈdʌktə(r)]	n.（乐队等的）指挥，检票员，导体
semiconductor	[ˌsemikənˈdʌktə(r)]	n. 半导体，半导体装置
insulator	[ˈɪnsjuleɪtə(r)]	n. 绝缘体
viscosity	[vɪˈskɒsəti]	n. 黏性，黏度

Notes

* Gently pressing the pipette against the bottom of the vessel while removing the rubber bulb will help prevent liquid from draining below the mark while you place your finger in place (alternatively, you may use an automatic aspirator that remains attached to the pipette).

在移开移液器的橡胶球时，将移液器轻轻抵在容器底部，这有助于在你的手指就位时，防止液体流出至标记线之下（或者，你可以使用连接在移液器上的自动抽吸器）。

Trivial — Last but not the Least

Physical properties of matter

The physical properties of matter are any properties that can be perceived or observed without changing the chemical identity of the sample. In contrast, chemical properties are those that can only be observed and measured by performing a chemical reaction that changes the molecular structure of the sample. Because physical properties encompass such a wide range of characteristics, they are further classified as either intense or extensive and either isotropic or anisotropic.

Intensive and extensive physical properties

Intensive physical properties do not depend on the size or mass of the sample. Examples of intensive properties are boiling point, state of matter, and density. Extensive physical properties depend on the amount of matter in the sample. Examples of extensive properties include size, mass, and volume.

Isotropic and anisotropic physical properties

Isotropic physical properties do not depend on the orientation of the sample or the direction from which it is viewed. Anisotropic properties depend on orientation. While any physical property can be considered isotropic or anisotropic, the terms are usually used to identify or distinguish materials based on their optical and mechanical properties. For example, one crystal may be isotropic in color and opacity, while another may appear different colors depending on the axis of observation. In a metal, grains may be distorted or elongated along one axis compared to another.

Physical properties of ionic vs. covalent bonds

The nature of the chemical bonds plays a role in some of the physical properties of a material. The ions in ionic compounds are strongly attracted to other ions of opposite charge and repelled by those of the same charge. Atoms in covalent molecules are stable and are not strongly attracted or repelled by other parts of the material. As a result, ionic solids tend to have higher melting and boiling points than the low melting and boiling points of covalent solids. Ionic compounds tend to be good electrical conductors when melted or dissolved, while covalent compounds tend to be poor conductors in any form. Ionic compounds are usually crystalline solids, while covalent molecules exist as liquids, gases, or solids. Ionic compounds often dissolve in water and other polar solvents, while covalent compounds tend to dissolve in nonpolar solvents.

Chemical properties

Chemical properties are characteristics of matter that can only be observed by changing the chemical identity of a sample — by studying its behavior in a chemical reaction. Examples of chemical properties include flammability (observed by burning), reactivity (measured by the willingness to participate in a reaction), and toxicity (demonstrated by exposing an organism to a chemical).

Chemical/physical property description

Chemical and physical properties are characteristics of matter that can be used to identify and describe it. Chemical properties are those that can only be observed when matter undergoes a chemical change or reaction. In other words, you must change the chemical identity of a sample in order to observe and measure its chemical properties. It's important to know the chemical properties of a sample because this information can be used to (1) classify the sample before we use it, (2) identify an unknown sample, (3) purify it, (4) separate it from other substances, (5) predict its behavior, and (6) predict its uses. Let's take a closer look at some examples of chemical properties.

Toxicity

Toxicity is an example of a chemical property. Toxicity is how dangerous a chemical is to your health, a specific organ, another organism, or the environment. You can't tell if a chemical is toxic or not just by looking at it. How toxic a substance is depends on the situation, so it is a property that can only be observed and measured by exposing an organic system to a sample.

The exposure causes a chemical reaction or series of reactions. The net result of the chemical changes is toxicity.

Flammability

Flammability is a measure of how easily a sample ignites or how well it can sustain a combustion reaction. You don't know how easily something burns until you try to ignite it, so flammability is an example of a chemical property. Note that flammable and ignitable are similar and both mean that something is combustible or can be ignited.

Chemical stability

Chemical stability is also known as thermodynamic stability. It occurs when a substance is in chemical equilibrium with its environment, which is its lowest energy state. This is a property of matter that is determined by its specific conditions, so it can't be observed without exposing a sample to that situation. Thus, chemical stability fits the definition of a chemical property of matter. Chemical stability is related to chemical reactivity. While chemical stability refers to a specific set of circumstances, reactivity is a measure of how likely a sample is to participate in a chemical reaction under a variety of conditions and how quickly a reaction might proceed.

Examples of description

Basically, a chemical property is a characteristic that can only be observed as a result of a chemical reaction. The following Table 2.1 contains some descriptive vocabulary related to physical and chemical properties.

Table 2.1 Descriptive vocabulary related to physical and chemical properties

color	colorless; red-brown; violet-black; purple-black; pale yellow; dark brown
state	solid; liquid; gas; gaseous; oily; crystalline; uncrystalline; molten; fused; melting point; boiling point
smell	odorless; pungent; penetrating; choking; sour; sweet; bitter
solubility	soluble; insoluble; slightly soluble; very soluble
observations	precipitate; milk; aqueous solution
density	heavy; light; less dense; about the same dense
hardness	hard; soft; ductile; malleable
toxicity	toxic; poisonous
conductivity	electronic conductivity; thermal conductivity; conductor; semiconductor; insulator
volume	large/small
temperature	high/low
pressure	hight/low
viscosity	hight/low
electrical charge	hight/low

Part B
Introduction to Fundamentals of Chemistry and Chemical Engineering

Unit 3

Basic Information of Inorganic Chemistry

3.1 Overview

Inorganic chemistry is defined as the study of the chemistry of materials of non-biological origin. Typically, this refers to materials that do not contain carbon-hydrogen bonds, including metals, salts, and minerals. Inorganic chemistry is used to study and develop catalysts, coatings, fuels, surfactants, materials, superconductors, and pharmaceuticals. Important chemical reactions in inorganic chemistry include double displacement reactions, acid-base reactions, and redox reactions. In contrast, the chemistry of compounds containing carbon-metal bonds is called organometallic chemistry, and overlaps both organic and inorganic chemistry. Organometallic compounds typically involve a metal bonded directly to a carbon atom. The key to all inorganic chemistry is the periodic table.

The first man-made inorganic compound of commercial importance to be synthesized was ammonium nitrate. Ammonium nitrate was produced by the Haber process for use as a soil fertilizer.

Properties of inorganic compounds

Because the class of inorganic compounds is so large, it's difficult to generalize about their properties. However, many inorganics are ionic compounds, containing cations and anions linked by ionic bonds. Classes of these salts include oxides, halides, sulfates, and carbonates. Another way to classify inorganic compounds is as main group compounds, coordination compounds, transition metal compounds, cluster compounds, organometallic compounds, solid state compounds and bioinorganic compounds[1]. Many inorganic compounds are poor electrical and thermal conductors as solids, have high melting points and readily adopt crystalline structures. Some are soluble in water, others are not. Usually, the positive and negative electrical charges balance to form neutral compounds. Inorganic chemicals are common in nature as minerals and electrolytes.

Inorganic chemists work in a wide range of fields. They may study materials, learn how to synthesize them, develop practical applications and products, and reduce the environmental impact of inorganic compounds. Examples of industries that employ inorganic chemists include government agencies, mining, electronics and chemical companies. Closely related fields include materials science and physics. Becoming an inorganic chemist generally requires a postgraduate degree (Master's or Ph.D.). Most inorganic chemists study chemistry at the university level.

Applications of inorganic chemistry

Inorganic chemistry is a wondrous subject in itself, with each element and compound having a unique property. But beyond its academic interest, inorganic chemistry is an integral part of the world we live in: from the air we breathe, to the water around us, to the rocks beneath our feet. Each of these subjects is essentially a branch of inorganic chemistry with its own name: atmospheric chemistry, aquatic chemistry and geochemistry. Living organisms on this planet depend on many chemical elements for their functioning, and this subject also has its own identity, that of bioinorganic chemistry. Nature is not the only context in which inorganic chemistry is applied. Our modern civilization depends on the synthesis of inorganic compounds, from the vast quantities of chemicals such as ammonia and sulfuric acid to the much smaller quantities of cadmium sulfide used as a paint pigment. Industrial inorganic chemistry is thus one of the cornerstones of our modern economy. But more than that, inorganic chemistry is part of our future, part of materials science, the synthesis of new and exotic materials for the 21st century that will enable us to live better, more comfortable and greener lives.

3.2 Four Types of Inorganic Chemical Reactions

Elements and compounds react with each other in a variety of ways. To memorize every type of reaction would be challenging and unnecessary, since almost every inorganic chemical reaction falls into one or more of four broad categories.

Combination reactions. Two or more reactants form a product in a combination reaction. An example of a combination reaction is the formation of sulfur dioxide when sulfur is burned in air:

$$S(s) + O_2(g) \longrightarrow SO_2(g)$$

Decomposition reactions. In a decomposition reaction, a compound is broken down into two or more substances. Decomposition usually results from electrolysis or heating. An example of a decomposition reaction is the breakdown of mercury(II) oxide into its component elements.

$$2HgO(s) \longrightarrow 2Hg(l) + O_2(g)$$

Single displacement reactions. A single displacement reaction is characterized by an atom or ion of a single compound replacing an atom of another element. An example of a single displacement reaction is the displacement of copper ions in a copper sulfate solution by zinc metal, forming zinc sulfate:

$$Zn(s) + CuSO_4(aq) \longrightarrow Cu(s) + ZnSO_4(aq)$$

Single displacement reactions are often subdivided into more specific categories (e.g., redox reactions).

Double displacement reactions. Double displacement reactions also be called metathesis reactions. In this type of reaction, elements from two compounds displace each other to form new compounds. Double displacement reactions may occur when one product is removed from the solution as a gas or precipitate or when two species combine to form a weak electrolyte that remains undissociated in solution[2]. An example of a double displacement reaction occurs when solutions of calcium chloride and silver nitrate are reacted to form insoluble silver chloride in a solution of calcium nitrate.

$$CaCl_2(aq) + 2\ AgNO_3(aq) \longrightarrow Ca(NO_3)_2(aq) + 2\ AgCl(s)$$

A neutralization reaction is a specific type of double displacement reaction that occurs when an acid reacts with a base, producing a solution of salt and water. An example of a neutralization reaction is the reaction of hydrochloric acid and sodium hydroxide to form sodium chloride and water:

$$HCl(aq) + NaOH(aq) \longrightarrow NaCl(aq) + H_2O(l)$$

Remember that reactions can belong to more than one category. Also, it would be possible to present more specific categories, such as combustion reactions or precipitation reactions. Learning the main categories will help you balance equations and predict the types of compounds formed from a chemical reaction.

3.3 The Important Inorganic Elements — Actinide

The "dual carbon targets" of "carbon peak" by 2030 and "carbon neutrality" by 2060 have been clearly proposed by the Chinese government in 2020. As a green energy, nuclear power has a very important role to play in this respect, where the zero or low carbon-based technology is even more desirable. In addition, the chemistry closely related to nuclear energy is radiochemistry, which refers to the chemistry of a special group, the actinide elements. There is a lot of work to be done in this area by chemists and chemical engineers.

The actinides are the 15 chemical elements with atomic numbers 89 through 103, the first member of which is actinium and the last member is lawrencium. The transactinides (those beyond the actinides) are the heaviest known chemical elements. Both the actinides and the transactinides have chemical properties that are governed by their outermost electronic subshells. Each of these groups of elements is a unique transition series (a group of elements in which d or f electronic subshells are being filled). The actinides are the transition elements that fill the 5f subshell. The actinide series is unique in several ways:

◆ Most of the elements (those heavier than uranium) were first discovered by synthetic methods: bombardment of heavy atoms with neutrons in nuclear reactors, bombardment with other particles in accelerators, or as the result of nuclear detonations.

◆ All actinide isotopes are radioactive, with a wide range of nuclear properties, especially that of spontaneous and induced nuclear fission.

◆ They are all metals with very large radii, and exist in chemical compounds and in solutions as cations with very large ionic radii.

◆ The metals exhibit an unusual range of physical properties. Plutonium, with six allotropes, is the most unusual of all metals.

◆ Many of the actinide elements have a large number of oxidation states. In this respect plutonium is unique, being able to exist in aqueous solution simultaneously in four oxidation states.

In metallic materials and in some other compounds containing elements lighter than plutonium, the 5f orbitals are sufficiently diffuse that the electrons in these orbitals are "itinerant" meaning they are delocalized and chemically bond, often resulting in unique magnetic moments and electrical conductivity. However, in most compounds with elements heavier than plutonium, the 5f electrons are "localized" and do not significantly contribute to electrical conductivity or chemical bonding. Materials containing plutonium and adjacent elements can exhibit both itinerant and localized behavior, depending on conditions such as temperature and applied pressure.

Actinium (which has no 5f electrons in the metal, free atom, or in any of its ions) and the elements americium through lawrencium are similar in many ways to the lanthanide elements (the elements that fill the 4f electron subshell). The elements thorium through neptunium have some properties similar to the d-transition elements.

Relativistic contributions to electronic properties and spin-orbit effects are important in the chemical properties of the actinides.

In the six decades since G. T. Seaborg's "actinide concept", great progress has been made in actinide and transactinide chemistry. New information and new concepts have accumulated to an extraordinary degree in these disciplines, and we refer, of course, to the problems posed by nuclear reactors for electricity generation, the production and dismantlement of nuclear weapons, the treatment and storage of nuclear waste, and the cleanup of Cold War nuclear material sites[3]. These are sources of acute global concern, all of which intimately involve the actinides.

The chemistry of the actinides has been remarkably well developed since the actinide concept itself was first publicly described in 1945. The elements thorium and uranium had already been studied by chemists for more than 100 years. Uranium enjoyed a small distinction as the heaviest element in nature, and as the terminus of the classical periodic table. In 1895, Becquerel had discovered that uranium underwent radioactive decay, a discovery that permanently removed uranium from its obscurity and ushered in the era of the Curies, Rutherford, Soddy, Hahn and Meitner, Fajans, and others who mapped the highly complex radioactive transformations of the naturally occurring elements[4]. The crucial importance of uranium, however, did not become fully apparent until Fermi and his colleagues irradiated many of the elements, including uranium, with neutrons in the 1930s. They produced new radioactive species with chemical properties that were not identical to any of the known heavy elements. The Fermi group believed that they had created new elements heavier than uranium. In 1938, Hahn, Meitner, and Strassmann performed definitive chemical experiments showing that the radioactive species produced by neutron irradiation of uranium were in fact fission fragments resulting from the splitting of the uranium nucleus into smaller nuclei. Their experiments constituted the discovery of nuclear fission. The earlier formation of transuranium elements had been disproved, but the way to their synthesis was now open.

New Words and Expressions

catalyst	[ˈkætəlɪst]	n. 催化剂
coating	[ˈkəʊtɪŋ]	n. 涂料，涂层，镀膜加工
fuel	[ˈfjuːəl]	n. 燃料；v. 供给燃料
surfactant	[sɜːrˈfæktənt]	n. 表面活性剂
superconductor	[ˈsuːpəkəndʌktə(r)]	n. 超导体
periodic	[ˌpɪəriˈɒdɪk]	adj. 阶段性的，定期的，（化学）元素周期表的，（修辞）完整句的

英文	音标	释义
synthesized	[ˈsɪnθəsaɪzd]	adj. 合成的，综合的；v. 合成（synthesize 的过去分词）
ammonium	[əˈməunɪəm]	n. 铵，氨盐基
nitrate	[ˈnaɪtreɪt]	n. 硝酸盐；v. 用硝酸处理
halide	[ˈhælaɪd]	n. 卤化物，卤化物类
sulphate	[ˈsʌlfeɪt]	n. 硫酸盐；v. 使与硫酸化合，硫酸盐化
carbonate	[ˈkɑːbənət]	n. 碳酸盐
mining	[ˈmaɪnɪŋ]	n. 采矿，采矿业
atmospheric	[ˌætməsˈferɪk]	adj. 大气（层）的，有关大气的，有神秘（美）感的，富有情调的
aquatic	[əˈkwætɪk]	adj. 水生的，水的；n. 水生植物（或动物）
civilization	[ˌsɪvəlaɪˈzeɪʃn]	n. 文明，文明社会，文明国家，人类社会，舒适的生活环境，教化，开化
cadmium	[ˈkædmɪəm]	n. 镉
pigment	[ˈpɪgmənt]	n. 色素，颜料；v. 给……染色，呈现颜色
decomposition	[ˌdiːˌkɒmpəˈzɪʃn]	n. 分解，腐烂，变质
mercury	[ˈmɜːkjəri]	n. 汞，水银（Mercury 水星）
chloride	[ˈklɔːraɪd]	n. 氯化物
actinide	[ˈæktɪnaɪd]	n. 锕类，锕系元素
fission	[ˈfɪʃn]	n. 裂变
radiochemistry	[ˌreɪdɪəuˈkemɪstri]	n. 放射化学
itinerant	[aɪˈtɪnərənt]	adj. 巡回的，流动的
lawrencium	[lɒˈrensɪəm]	n. 铹（人工获得的放射性元素，超铀元素之一）
subshell	[ˈsʌbʃel]	n. 支壳层
bombardment	[bɒmˈbɑːdmənt]	n. 轰炸，炮击
plutonium	[pluːˈtəunɪəm]	n. 钚
simultaneously	[ˌsɪm(ə)lˈteɪnɪəsli]	adv. 同时地
orbital	[ˈɔːbɪtl]	n. 轨道
magnetic	[mægˈnetɪk]	adj. 磁的，磁性的，磁化的，有吸引力的
neptunium	[nepˈtjuːnɪəm]	n. 镎
uranium	[juˈreɪnɪəm]	n. 铀
relativistic	[ˌrelətɪˈvɪstɪk]	adj. 相对论的，相对的
dismantlement	[dɪsˈmæntlmənt]	n. 拆卸，拆除的行动或状态
terminus	[ˈtɜːmɪnəs]	n. 终点，终点站
obscurity	[əbˈskjuərəti]	n. 默默无闻，晦涩，无名
usher	[ˈʌʃə(r)]	v. 引领，引导，使开始

Notes

1. Another way of classifying inorganic compounds is as main group compounds, coordination compounds, transition metal compounds, cluster compounds, organometallic compounds, solid state compounds and bioinorganic compounds.

无机化合物的另一种划分是主基团化合物、配位化合物、过渡金属化合物、簇状化合物、有机金属化合物、固态化合物和生物无机化合物。

2. Double displacement reactions may occur when one product is removed from the solution as a gas or precipitate or when two species combine to form a weak electrolyte that remains undissociated in solution.

当一种产物以气体或沉淀物的形式从溶液中除去，或者当两种物质结合在溶液中形成未解离的弱电解质时，可能发生双置换反应。

3. New information and new concepts have accumulated to an extraordinary degree in these disciplines, and we refer, of course, to the problems posed by nuclear reactors for electricity generation, the production and dismantlement of nuclear weapons, the treatment and storage of nuclear waste, and the cleanup of Cold War nuclear material sites.

在这些学科中，新的信息和新的概念已经积累到了非同寻常的程度，当然，我们指的是发电用核反应堆、核武器的生产和拆卸、核废料的处理和储存以及冷战时期核材料场地的清理所带来的问题。

4. In 1895, Becquerel had discovered that uranium underwent radioactive decay, a discovery that permanently removed uranium from its obscurity and ushered in the era of the Curies, Rutherford, Soddy, Hahn and Meitner, Fajans, and others who mapped the highly complex radioactive transformations of the naturally occurring elements.

1895年，贝克勒尔发现铀发生了放射性衰变，这一发现将默默无闻的铀变成了众人瞩目的焦点，并开创了居里夫妇、卢瑟福、索迪、哈恩和迈特纳、法詹斯等科学家绘制自然元素高度复杂的放射性转化图的时代。

Trivial — Last but not the Least

General chemistry descriptions and expressions

1. 怎样读化学方程式？以下是几个读方程式的案例：

Example 1　　　　　　　　　　　　　　　　$N_2 + 3H_2 \xrightleftharpoons[\text{催化剂}]{\text{高温、高压}} 2NH_3$

(1) Nitrogen reacts with hydrogen to form ammonia at high temperature and pressure with the presence of a catalyst.

(2) 1 mol nitrogen reacts with 3 mol hydrogen to form 2 mol ammonia at high temperature and pressure with the presence of a catalyst.

(3) Reaction between nitrogen and hydrogen at high temperature and pressure with the presence of a catalyst gives ammonia.

(4) At high temperature and pressure, reaction of nitrogen with hydrogen in the presence of a catalyst takes place (occurs).

Example 2　　　　　　　　　　　　　　　　$Zn + 2HCl = ZnCl_2 + H_2\uparrow$

zinc treated with hydrochloric acid forms hydrogen and zinc chloride.

Example 3　　　　　　　　　　　　　　　　$CaCO_3 \xrightarrow{\triangle} CaO + CO_2\uparrow$

(1) When calcium carbonate heated produces calcium oxide and carbon dioxide.

(2) Calcium carbonate is heated to yield calcium oxide and carbon dioxide.

(3) Calcium carbonate decomposes to calcium oxide and carbon dioxide when it is heated.

---

2. 有关化学计算过程中涉及的英文表述如下：
(1) 四则混合运算相关读法和写法

项目	四则混合运算的英文表述			
运算符号	+	−	×	÷
运算名称	addition	substraction	multiplication	division
动词读法	add	substract(ed) from	multiply(ied) by	divide(d) by
介词读法	plus	minus	times	over
运算结果	sum	difference	product	quotient

(2) 常见数学运算相关读法与写法

写法	读法
0.001	o point o o one / zero point zero zero one
$2/3$	two thirds
=/≡	equals / is equal to
≈	is approximately equal to
<	less than
>	greater than
x^2	x squared
x^3	x cubed
x^{-10}	x to the minus tenth power
100℃(℉)	one hundred degrees centigrade (Celsius) / 100 degrees Fahrenheit
5%	five percent
{ }	braces
/	per
∝	be proportional to
$\log_n X$	log X to the base n
$\log_{10} X$	log X to the base 10 (common logarithm)
$\log_e X$	log X to the base e (natural logarithm)
b_1	b sub 1
$F(x)$	function F of x
b', b'', b'''	b prime, b double prime, b triple prime
dx	differential x
$dy/dx, d^n y/dx^n$	the first derivative of y with respect to x, the n-th derivative of y with respect to x
$\partial y/\partial u$	the partial derivative of y with respect to u
x^n	the n-th power of x / x to the power n

(3) 部分常用形状的描述如下:

Planar		Steroscopic	
Linear	——	pyramid	

Part B Introduction to Fundamentals of Chemistry and Chemical Engineering   043

(continued)

Planar		Steroscopic	
triangle	△	tetrahedron	(tetrahedron, 109.5°)
square	□	sphere	(sphere)
rectangle	▭	tetragonal bipyramid	(tetragonal bipyramid)
hexagon	⬡	hexagonal pyramid	(hexagonal pyramid)
oval	⬭	triangular prism	(triangular prism)
circular	○	hexagonal prism	(hexagonal prism)

Unit 4
Rules of Inorganic Nomenclature

In general, there are two types of inorganic compounds that can be formed: ionic compounds and molecular compounds. Nomenclature is the process of giving chemical compounds different names so that they can be easily identified as separate chemicals. Inorganic compounds are compounds that are not involved in the formation of carbohydrates, or simply any compound that does not fit into the description of an organic compound. For example, organic compounds include molecules with carbon rings and/or chains of hydrogen atoms. Inorganic compounds, the subject of this unit, are any molecule that does not have these characteristic carbon and hydrogen structures.

The standards for nomenclature in chemistry are the rules published by the International Union of Pure and Applied Chemistry (IUPAC). The rules should not be regarded as a rigid code, but as an evolving attempt to clarify the process of naming. It should be recognized that one purpose of any set of rules is to prevent the proliferation of non-systematic names that have been invented and/or used for certain compounds. Nevertheless, many trivial names will appear in common usage. For example, it is not expected that ethanoate will soon replace acetate for the anion CH_3COO^-. The following is intended as a brief outline to acquaint the reader with the general principles of inorganic nomenclature.

4.1 Oxidation Number

The concept of oxidation number is interwoven in the fabric of inorganic chemistry in many ways, including nomenclature. The oxidation number of an element in a chemical compound is the charge that would be present on an atom of the element if the electrons in each bond were assigned to the more electronegative atom[1].

	Element	Oxidation No.	Element	Oxidation No.
MnO_4^- (one Mn^{7+} and four O^{2-} ions)	Mn	VII	O	$-$ II
ClO^- (one Cl^+ and one O^{2-} ion)	Cl	I	O	$-$ II

4.2 Coordination Number

The coordination number of the central atom in a compound is the number of atoms directly attached to the central atom. The attached atoms may be charged, uncharged, part of an ion, or part of a molecule. In certain coordination compounds, one coordination position is assigned to two atoms of a multiple bond within an attached group. The coordination number of an atom or ion in a lattice is defined by crystallographers as the number of close neighbors in this arrangement.

4.3 Use of Multiplicative Prefixes, Enclosing Marks, Numbers, and Letters

Chemical nomenclature uses multiplicative prefixes, numbers (both Arabic and Roman), and letters to indicate both stoichiometry and structure.

4.3.1 Multiplicative Prefixes

The simple multiplicative prefixes mono, di, tri, tetra, penta, hexa, octa, nona (ennea), deca, undeca (hendeca), dodeca, etc. indicate the following situations:

(1) Stoichiometric proportions:

CO	carbon monoxide
CO_2	carbon dioxide

(2) Extent of substitution:

$SiCl_2H_2$	dichlorosilane
$PO_2S_2^{2-}$	dithiophosphate ion

(3) Number of identical coordinated groups:

$[Co(NH_3)_4Cl_2]^+$	tetraamminedichlorocobalt(III) ion

The prefixes are used in slightly different ways to denote (1) the number of identical central atoms in condensed acids and their characteristic anions, or (2) the number of atoms of the same element that form the skeleton of some molecules or ions:

H_3PO_4	(mono)phosphoric acid ("mono" should be omitted before the first element)
$H_4P_2O_7$	diphosphoric acid
$H_2S_3O_{10}$	trisulfuric acid
Si_2H_6	disilane
$S_4O_6^{2-}$	tetrathionate ion

The multiplicative prefixes bis, tris, tetrakis, pentakis, etc. were originally introduced into organic nomenclature to denote a set of identical radicals, each substituted in the same way, and are sometimes used to avoid ambiguity[2]. The "tris(decyl)" in the second example below avoids any ambiguity with the organic radical tridecyl, $C_{13}H_{27}$. The use of these prefixes has been extended to avoid any misunderstanding, as shown in the fourth case below, the ligand is clearly CH_3NC.

$(ClCH_2CH_2)_2NH$	bis(2-chloroethyl) amine
$P(C_{10}H_{21})_3$	tris(decyl)phosphine
$[P(CH_2OH)_4]^+Cl^-$	tetrakis(hydroxymethyl)phosphonium chloride
$[Fe(CN)_2(CH_3NC)_4]$	dicyanotetrakis (methyl isocyanide) iron(II)

4.3.2 Enclosing Marks

Parentheses are used in formulae to surround sets of identical atoms, for instance, $Ca_3(PO_4)_2$ and $B[N(CH_3)_2]_3$. In names, parentheses are generally used after bis, tris, etc., around all complex expressions, and elsewhere to avoid any possibility of ambiguity. In the formulae of coordination compounds, square brackets are used to enclose a complex ion or neutral coordination entity, for example, $K_3[Co(C_2O_4)_3]$ and $[Co(NH_3)_3(NO_2)_3]$.

4.3.3 Numbers

In the names of inorganic compounds, Arabic numerals are used to denote the atoms on which there is a substituent or addition in a chain, for example, $H_3Si—ClSiH—SiH_2SiH_3$ is 2-chlorotetrasilane. This system is similar to that used in organic nomenclature and will not be discussed further here.

Arabic numerals enclosed in parentheses with a positive (+) or negative (−) symbol indicate the charge of the central ion in parent coordinated entities. The name of an uncharged coordination compound does not include a zero. Arabic numerals are sometimes used in place of multiplicative affixes. Roman numerals are used in brackets to indicate the oxidation number of an element.

$AlK(SO_4)_2·12H_2O$ aluminum potassium sulfate 12-water (in place of aluminum potassium sulfate dodecahydrate)

B_6H_{10} hexaborane(10) (in place of hexaboron decahydride)

4.3.4 Italic Letters

The symbols in italics are used to indicate the following situations:
(1) the element in a heteroatomic chain or ring on substitution occurs:
CH_3ONH_2 *O*-methylhydroxylamine
(2) the element in a ligand that is coordinated to the central atom:

cysteinato-*S*, *N*…

(3) the presence of bonds between two metal atoms:
$(OC)_3Fe(C_2H_5S)_2Fe(CO)_3$ bis(*μ*-ethylthio)bis(tricarbonyliron)(Fe-Fe)
(4) the point of attachment in some addition compounds:
$CH_3ONH_2BH_3$ *O*-methylhydroxylamine(N-B)borane

Although there is a conventional treatment for italicizing fonts for naming compounds, recent literature reports have described compounds using orthographic fonts that comply with journal requirements, rather than following the aforementioned rules. Thus, it is inappropriate to generalize the use of italicized fonts.

4.4 Elements

A chemical **element**, or element, is defined as any substance that cannot be decomposed into simpler substances by ordinary chemical processes. Elements are the fundamental materials of which all matter is composed. There are 118 known elements. Each element is identified by the number of protons in its atomic nucleus. A new element can be created by adding more protons to an atom.

Atoms of the same element have the same atomic number, or *Z*. Each element can be represented by its atomic number or by its atomic symbol. The element symbol is an abbreviation of one or two letters. The first letter of an element symbol is always capitalized. A second letter, if present, is written in lower case. The International Union of Pure and Applied Chemistry (IUPAC) has agreed on a set of names and symbols for the elements that are used in the scientific literature. However, the names and symbols for the elements may be different in different countries. For example, element 56 is called barium by IUPAC and in English with the symbol Ba. It is called bario in Italian and baryum in French. The element with atomic number 4 is called boron by IUPAC, but boro in Italian, Portuguese, and Spanish, bor in German, and bore in French. Common element symbols are used by countries with similar alphabets.

When the atomic or ionic symbol of an element is recorded, the mass number, atomic number, number of atoms, and ionic charge of an element are indicated by means of four indices placed around the symbol:

$$\text{mass number} \quad \text{ionic charge}$$
$$\text{SYMBOL} \qquad {}^{15}_{7}N^{3-}_{2}$$
$$\text{atomic number} \quad \text{number of atoms}$$

Ionic charge should be indicated by an Arabic superscript numeral preceding the plus or minus sign: Mg^{2+}, PO_4^{3-}, etc.

Element synthesis. Atoms of an element can be produced by the processes of fusion, fission and radioactive decay. These are all nuclear processes, which means that they involve the protons and neutrons in the nucleus of an atom. In contrast, chemical processes (reactions) involve electrons rather than nuclei. In fusion, two atomic nuclei combine to form a heavier element. In fission, heavy nuclei split to form one or more lighter nuclei. Radioactive decay can produce different isotopes of the same element or a lighter element. When the term "chemical element" is used, it can refer to a single atom of that element, or to any pure substance consisting only of that type of element. For example, an atom of iron and a bar of iron are both elements of the chemical element iron.

The periodic table, also known as the table of the elements, is an organized arrangement of the 118 known chemical elements. The chemical elements are arranged from left to right and from top to bottom in order of increasing atomic number, or the number of protons in an atomic nucleus, which generally coincides with increasing atomic mass.

4.5 Formulae and Names of Compounds in General

Compounds of a metal and a nonmetal are commonly known as **ionic compounds**, where the compound name ends with -ide. Cations have a positive charge and anions have a negative charge. The net charge of an ionic compound must be zero, which also means that it is electrically neutral. For example, one Na^+ is paired with one Cl^-, one Ca^{2+} is paired with two Br^-. There are two rules to follow: (1) the cation (metal) is always written first, with its name unchanged; and (2) the anion (non-metal) is written after the cation, modified to end in -ide. In formulae, the electropositive component (cation) should always be written first, e.g., KCl, $CaSO_4$. In the case of binary compounds between non-metals (those containing atoms of two elements), they are named stoichiometrically by combining the element names and, by convention, treating the first element reached by following the arrow in the element sequence below (Figure 4.1) as if it were an anion, e.g., XeF_2, NH_3, H_2S, S_2Cl_2, Cl_2O, OF_2.

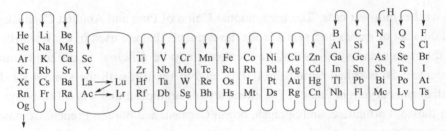

Figure 4.1 Element sequence

So the name of this formally "electronegative" element is given an "ide" ending and placed after the name of the formally "electropositive" element, followed by a space. For binary compounds, the name of the element that comes later in the sequence is modified to end in -ide. If the electronegative constituent is heteropolyatomic, it should be given the suffix -ate. In certain exceptional cases, the suffixes -ide and -ite are used. The transition metals may form more than one ion, thus it is necessary to specify which particular ion we are talking about. This is indicated by assigning a Roman numeral after the metal. The Roman numeral denotes the charge and the oxidation state of the transition metal ion. For example, iron can form two common ions, Fe^{2+} and Fe^{3+}. To distinguish the difference, Fe^{2+} would be named iron(Ⅱ) and Fe^{3+} would be named iron(Ⅲ).

Chemical	Name	Chemical	Name
CaF_2	calcium (di)fluoride	H_2O_2	dihydrogen dioxide / hydrogen peroxide
$FeCl_2$	iron dichloride / iron(Ⅱ) chloride	$FeCl_3$	iron trichloride / iron(Ⅲ) chloride
CO_2	carbon dioxide	GaAs	gallium arsenide
NaCl	sodium chloride	CaS	calcium sulfide
As_2Se_3	arsenic selenide	Li_3N	lithium nitride
H_2S	hydrogen sulfide	$NiAs_2$	nickel arsenide
SiC	silicon carbide	HCl	hydrogen chloride
OF_2	oxygen difluoride	CS_2	carbon disulfide
ClO_2	chlorine dioxide	SF_6	sulfur hexafluoride

Compounds that consist of a nonmetal bonded to a nonmetal are commonly known as **molecular compounds**, where the element with the positive oxidation state is listed first. In many cases, nonmetals form more than one binary compound, so prefixes are used to distinguish them. The prefix mono- is not used for the first element. If there is no prefix before the first element, it is assumed that there is only one atom of that element.

In inorganic compounds, it is generally possible to identify a characteristic atom (as in ClO^-) or a central atom (as in ICl_4^-) in a polyatomic group. Such a polyatomic group is called a complex, and the atoms, radicals or molecules attached to the characteristic or central atom are called ligands. In this case, the name of the negatively charged complex should be formed from the name of the characteristic or central element modified to end in -ate.

Binary hydrogen compounds may be named according to the principles. Volatile hydrides, other than those of Group Ⅶ and of oxygen and nitrogen, may also be named by giving the stem name of the element followed by the suffix -ane. If the molecule contains more than one atom of that element, the number is indicated by the appropriate Greek prefix.

Recognized exceptions are water, ammonia and hydrazine due to their long history of use. Phosphine, arsine, stibine and bismuthine are also acceptable. However, "-ane" names should be used for all molecular hydrides containing more than one atom of the element.

Chemical formula	Name	Chemical formula	Name
B_2H_6	diborane	PbH_4	plumbane
Si_3H_8	trisilane	H_2S_n	polysulfane

4.6 Names for Ions and Radicals

Monatomic cations should be named as the corresponding element, without modification or suffix. However, some of the transition metal charges have specific Latin names. Just like some conventional naming rules, the transition metal ion with the lower charge has a Latin name ending in -ous, and the one with the higher charge has a Latin name ending in -ic. The most common ones are listed below:

Ion	Name	Ion	Name
Cu^+	copper(I) (cat)ion / cuprous	Cu^{2+}	copper(II) (cat)ion / cupric
Fe^{2+}	iron(II) (cat)ion / ferrous	Fe^{3+}	iron(III) (cat)ion / ferric
Hg_2^{2+}	mercury(I) (cat)ion / mercurous	Hg^{2+}	mercury(II) (cat)ion / mercuric
Pb^{2+}	lead(II) (cat)ion / plumbous	Pb^{4+}	lead(IV) (cat)ion / plumbic
Sn^{2+}	tin(II) (cat)ion / stannous	Sn^{4+}	tin(IV) (cat)ion / stannic
I^+	iodine(I) cation	I^-	iodine(I) ion / iodine ion

The above principle should also apply to polyatomic cations corresponding to radicals for which specific names are given and these names should be used without modification or addition, e.g., the nitrosyl cation (NO^+), the nitryl cation (NO_2^+).

Ion	Name	Ion	Name
H^-	hydride ion	N^{3-}	nitride ion
F^-	fluoride ion	O^{2-}	oxide ion

In **polyatomic** ions, polyatoms (i.e., two or more atoms) are bonded together by covalent bonds. Although there may be an element with a positive charge, such as H^+, it is not bonded to another element with an ionic bond. This occurs because if the atoms were to form an ionic bond, they would have already formed a compound and would not need to gain or lose any electrons. Polyatomic anions are more common than polyatomic cations, as shown below. Polyatomic anions have negative charges while polyatomic cations have positive charges. To indicate different polyatomic ions made up of the same elements, the name of the ion is modified according to the following example:

Ion	Name	Ion	Name
NH_4^+	ammonium ion	CN^-	cyanide ion
H_3O^+	hydronium ion	OH^-	hydroxide ion
$C_2H_3O_2^-$	acetate ion	NO_2^-	nitrite ion
AsO_4^{3-}	arsenate ion	NO_3^-	nitrate ion

Formula	Name	Formula	Name
CO_3^{2-}	carbonate ion	$C_2O_4^{2-}$	oxalate ion
ClO^-	hypochlorite ion	MnO_4^-	permanganate ion
ClO_2^-	chlorite ion	PO_4^{3-}	phosphate ion
ClO_3^-	chlorate ion	SO_3^{2-}	sulfite ion
ClO_4^-	perchlorate ion	SO_4^{2-}	sulfate ion
CrO_4^{2-}	chromate ion	SCN^-	thiocyanate ion
$Cr_2O_7^{2-}$	dichromate ion	$S_2O_3^{2-}$	thiosulfate ion

Polyatomic cations formed by adding extra protons to monatomic anions to achieve a neutral charge are named with the -onium suffix attached to the root of the anion element's name. Examples of such cations include phosphonium, arsonium, sulfonium, and iodonium. The name ammonium for the ion NH_4^+ does not follow the above rule but is retained. Substituted ammonium ions derived from nitrogen bases with names ending in -amine are given names formed by changing -amine to -ammonium. For example, $HONH_3^+$ is the hydroxylammonium ion. If the nitrogen base is known by a name other than -amine, the cation name is formed by adding -ium to the name of the base (omitting a final e or other vowel if necessary): hydrazinium, anilinium, glycinium, pyridinium. The names of monatomic anions consist of the names (sometimes abbreviated) of the elements ending in -ide, and some polyatomic anions also have names ending in -ide.

Chemical formula	Name	Chemical formula	Name
OH^-	hydroxide ion	N_3^-	azide ion
O_2^{2-}	peroxide ion	NH^{2-}	imide ion
S_2^{2-}	disulfide ion	NH_2^-	amide ion
I_3^-	triiodide ion	CN^-	cyanide ion
HF_2^-	hydrogen difluoride ion	C_2^{2-}	acetylide ion

Treating oxygen as a typical ligand is practical, although traditionally, its name has been omitted when indicating its presence and proportion in anions. Instead, a series of prefixes and occasionally the suffix -ite are used in place of -ate. The suffix -ite has been used to denote a lower oxidation state and may be retained in trivial names in the following cases:

Chemical formula	Name	Chemical formula	Name
NO_2^-	nitrite	$S_2O_7^{2-}$	pyrosulfate
$N_2O_2^{2-}$	hyponitrite	$S_2O_8^{2-}$	persulfate
NOO_2^{2-}	peroxonitrite	$S_4O_6^{2-}$	tetrathionate
AsO_3^{3-}	arsenite	SeO_3^{2-}	selenite
SO_3^{2-}	sulfite	ClO_2^-	chlorite
$S_2O_4^{2-}$	dithionite	ClO^-	hypochlorite
$S_2O_5^{2-}$	metabisulfite	BrO^-	hypobromite
$S_2O_6^{2-}$	dithionate	IO^-	hypoiodite

A radical is considered here to be a group of atoms which occurs repeatedly in a number of different compounds. Certain neutral and cationic radicals containing oxygen or other chalcogens, regardless of their charge, have special names ending in -yl, and the Commission on the Nomenclature of Inorganic Chemistry has agreed to retain the following for the time being:

Part B Introduction to Fundamentals of Chemistry and Chemical Engineering   051

Chemical formula	Name	Chemical formula	Name
HO^-	hydroxyl	SO^-	sulfinyl (thionyl)
CO^-	carbonyl	SO_2^-	sulfonyl (sulfuryl)
NO^-	nitrosyl	$S_2O_5^-$	disulfuryl
NO_2^-	nitryl	SeO^-	selenenyl
PO^-	phosphoryl	SeO_2^-	selenonyl
ClO^-	chlorosyl	CrO_2^-	chromyl
ClO_2^-	chloryl	UO_2^-	uranyl
ClO_3^-	perchloryl	NpO_2^-	neptunyl

Radicals analogous to the above, containing other chalcogens instead of oxygen, are designated by adding the prefixes thio-, seleno-, etc. These polyatomic radicals are always treated as forming the positive part of the compound:

$COCl_2^-$ carbonyl chloride
$PSCl_3^-$ thiophosphoryl chloride
SO_2NH^- sulfonyl (sulfuryl) imide

In the past, the names sulfinyl and sulfonyl chloride were more popular with organic chemists and the names thionyl and sulfuryl were more popular with inorganic chemists. It should be noted that the same radical may have different names in inorganic and organic chemistry. For example, the names of purely organic compounds, such as organic ligands, should follow the nomenclature of organic chemistry. The nomenclature of organic chemistry is largely based on the scheme of substitution, i.e. the replacement of hydrogen atoms by other atoms or groups. Such "substitutive names" are extremely rare in inorganic chemistry and are used in the following cases: NH_2Cl is called chloramine and $NHCl_2$ is called dichloramine. Other substitutive names include fluoro and chlorosulfonic acid, etc. These names should be replaced as follows.

HSO_3F fluorosulfuric acid
HSO_3Cl chlorosulfuric acid
NH_2SO_3H amidosulfuric acid (often referred to as "sulfamic acid")

4.7 Inorganic Acids

In general, the formulae of inorganic acids have H written first, whereas hydrogen-containing inorganic compounds that are not acids do not have H written first, e.g. HCl, HCN, HNO_3, H_2SO_4, H_3PO_4, etc., are acids and LiH, BeH_2, NH_3, PH_3, etc., are not acids (they are hydrides). There are two types of inorganic acids, those that contain oxygen (oxoacid) and those that do not (binary acid), and they are named differently. Always say "acid" when naming an acid.

4.7.1 Oxoacid

An **oxoacid** is an acid that contains hydrogen, oxygen, and another element. The anion formed when an oxoacid dissolves in water is an oxoanion. The names of oxoacids follow the rules:

(1) If a central atom can form two different oxoanions, the one with more oxygen atoms is named with the -ate suffix, and the one with fewer oxygen atoms is named with the -ite suffix, as discussed in 4.6 Names for Ions and Radicals.

Anion formula	Name	Anion formula	Name
NO_3^-	nitrate ion	NO_2^-	nitrite ion
SO_4^{2-}	sulfate ion	SO_3^{2-}	sulfite ion

(2) If the anion name ends in -ate, the corresponding acid name ends in -ic.

Anion formula	Name	Acid formula	Name
NO_3^-	nitrate ion	HNO_3	nitric acid
SO_4^{2-}	sulfate ion	H_2SO_4	sulfuric acid

(3) If the anion name ends in -ite, the corresponding acid name ends in -ous.

Anion formula	Name	Acid formula	Name
NO_2^-	nitrite ion	HNO_2	nitrous acid
SO_3^{2-}	sulfite ion	H_2SO_3	sulfurous acid

(4) When a central atom can form three or four oxoacids, both the anions and the oxoacids are distinguished by the prefixes hypo- and per- for the species with the fewest and most oxygen atoms, respectively. Note that the prefix per- is commonly used to designate a higher oxidation state and is retained for $HClO_4$, perchloric acid, and corresponding acids of other Group VII elements, and this use of the prefix per- should not be extended to elements of other groups.

Anion formula	Name	Acid formula	Name
ClO^-	hypochlorite ion	$HClO$	hypochlorous acid
ClO_2^-	chlorite ion	$HClO_2$	chlorous acid
ClO_3^-	chlorate ion	$HClO_3$	chloric acid
ClO_4^-	perchlorate ion	$HClO_4$	perchloric acid

The prefixes ortho-, meta-, and pyro- are used to distinguish acids that differ in their water content.

Acid formula	Name	Acid formula	Name
H_3BO_3	orthoboric acid	$(HBO_2)_n$	metaboric acid
H_4SiO_4	orthosilicic acid	$(H_2SiO_3)_n$	metasilicic acid
H_3PO_4	orthophosphoric acid / phosphoric acid	$(HPO_3)_n$	metaphosphoric acid
		$H_4P_2O_7$	pyrophosphoric acid / diphosphoric acid
H_6TeO_6	orthotelluric acid		
H_5IO_6	orthoperiodic acid	$(HO)_2OP-PO(OH)_2$	hypophosphoric acid / diphosphoric(IV) acid
$H_5P_3O_{10}$	triphosphoric acid		

Note that the removal of water has resulted in polymerization to form chains or rings

4.7.2 Acid Anions

Acid anions have H atoms that they can lose as H^+ in water. The names of these ions add hydrogen to the name of the corresponding ion that does not have H in it. If the acid anion has two or more hydrogen atoms capable to forming H^+, the appropriate Greek prefix is used to indicate the number. Mono- is omitted if only one acid anion is possible.

Acid ion	Name	Acid ion	Name
HCO_3^-	hydrogen carbonate ion	HPO_4^{2-}	monohydrogen phosphate ion
HSO_4^-	hydrogen sulfate ion	$H_2PO_4^-$	dihydrogen phosphate ion

4.7.3 Binary Hydrogen Compounds (Binary Acid)

Binary hydrogen compounds with nonmetals can form H^+ ion and an anion when dissolved in water. The acidic solutions are named as if they were molecular acids, using the usual name for the compound itself, replacing hydrogen with hydro- and the suffix -ide with -ic. The word acid is then added ("hydro" + element root + "ic" + acid). The formula for such a compound in water is often distinguished from the compound itself by (aq), indicating an aqueous solution.

Compound	Name	Acid solution	Name
HCl	hydrogen chloride	HCl(aq)	hydrochloric acid
HCN	hydrogen cyanide	HCN(aq)	hydrocyanic acid
H_2S	hydrogen sulfide	H_2S(aq)	hydrosulfuric acid

HCN, although not a binary compound, is analogous to the binary hydrogen halides (HCl, HBr, HI), and so as an acid is named in a similar manner. The name of H_2S(aq) as an acid is slightly irregular in using the full stem name of the element.

4.8 Salts

4.8.1 Simple Salts

Simple salts fall under the broad definition of binary compounds given in 4.5 Formulae and Names of Compounds in General, and their names are formed from those of the constituent ions in the manner set out in 4.6 Names for Ions and Radicals.

4.8.2 Salts Containing Acid Hydrogen

Names are formed by adding the word "hydrogen" with a numerical prefix, if necessary, to denote the displaceable hydrogen in the salt. The word hydrogen is placed immediately before the name of the anion. It should be noted that, in addition to the above rule, the prefix "bi-" may be added to the names of the HCO_3^-, HSO_3^-, and HSO_4^- ions before their original acid radical names.

$NaHCO_3$	sodium hydrogencarbonate / sodium bicarbonate
LiH_2PO_4	lithium dihydrogenphosphate
Na_2HPO_4	disodium hydrogenphosphate
KHS	potassium hydrogensulfide
$NaHSO_3$	sodium hydrogensulfite / sodium bisulfite

4.8.3 Hydrated Salts

Some ionic compounds (salts) contain water in their crystal structure and these compounds are called "hydrated salts". The added water changes their formula in a regular and measurable way, and knowing the amount of hydration of the salt is important in calculating its presence, as it is part of the mass. When naming hydrated salts, please note that the salt part is named according to the conventional method, while the number of molecules containing water of crystallization is expressed with a Greek prefix, i.e. the naming method is: salt name + Greek prefix (number) + hydrate.

In addition, salts containing water of crystallization may also be named according to the rule for naming addition compounds given in 4.10 Addition Compounds, where the names of the salts and the numbers of water molecules (or ratio) are connected by hyphens, with water always listed last in this order.

$CuSO_4 \cdot 5H_2O$	copper(II) sulfate pentahydrate
$Cr(NO_3)_3 \cdot 9H_2O$	chromium(III) nitrate nonahydrate
$BaCl_2 \cdot 2H_2O$	barium chloride dihydrate
$MnSO_4 \cdot H_2O$	manganese(II) sulfate monohydrate
$LiClO_4 \cdot 3H_2O$	lithium perchlorate trihydrate
$(NH_4)_2Mg(SO_4)_2 \cdot 6H_2O$	ammonium magnesium sulfate hexahydrate
$KAl(SO_4)_2 \cdot 12H_2O$	potassium aluminum sulfate dodecahydrate/ aluminum sulfate-potassium sulfate-water (1/1/24)
$Na_2CO_3 \cdot 10H_2O$	sodium carbonate decahydrate
	sodium carbonate 10-water
	sodium carbonate-water (1/10)

4.9 Coordination Compounds

Coordination compounds are those in which the central metal atom is linked by coordinate bonds to a number of ions or neutral molecules, either by these ions or neutral molecules donating lone pairs of electrons to the central metal atom. In formulae, it is usual to place the symbol for the central atom(s) first (except in formulae which are primarily structural), followed by the ionic and neutral ligands, and the formulae for the whole complex in square brackets. In the names, the central atom(s) should be placed after the ligands.

4.9.1 Indication of Oxidation Number and Proportion of Constituents

The nomenclature of coordination units is designed to denote the charge of the central atom/ion from which the unit originates. Since the charge of the coordination entity is the algebraic sum of the charges of its constituents, the necessary information may be provided by giving either the Stock number (formal charge of the central ion, i.e., oxidation number) or the Ewens-Bassett number[3].

$K_3[Fe(CN)_6]$	potassium hexacyanoferrate(III)
	potassium hexacyanoferrate(3−)
	tripotassium hexacyanoferrate
$K_4[Fe(CN)_6]$	potassium hexacyanoferrate(II)
	potassium hexacyanoferrate(4−)
	tetrapotassium hexacyanoferrate

Structural information may be indicated in formulae and names by prefixes such as *cis*, *trans*, *fac*, *mer*, etc. Anions are given the suffix -ate and cations and/or neutral molecules are given without a distinguishing ending. Ligands are listed in alphabetical order, regardless of their number of each. The name of a ligand is treated as a unit. Thus, "diamine" is listed under "a" and "dimethylamine" is listed under "d" [until the current rules were published in 1971, this rule required that the negative ligands to be listed first, e.g. dichlorodiammine platinum(II) rather than the suggested diamminedichloroplatinum(II). The older usage is found in much of the cited literature].

4.9.2 The Names of Anionic Ligands, whether Inorganic or Organic, End in -o

In general, when the anion name ends in -ide, -ite, or -ate, the last -e is replaced by -o, giving -ido, -ito, and -ato, respectively. Inorganic anionic ligands with numerical prefixes, such as (triphosphato), and thio, seleno, and telluro analogues of oxo anions containing more than one atom, such as (thiosulfate), must be preceded by a suffix. Examples of organic anionic ligands so named are:

Ion	Name	Ion	Name
CH_3COO^-	acetato	$(CH_3)_2N^-$	dimethylamido

The anions listed below do not follow the above rule exactly, and modified forms have become established:

Symbol	Ion	Ligand	Symbol	Ion	Ligand
F^-	fluoride	fluoro	H^-	hydride	hydrido/hydro*
Cl^-	chloride	chloro	OH^-	hydroxide	hydroxo
Br^-	bromide	bromo	O_2^{2-}	peroxide	peroxo
I^-	iodide	iodo	CN^-	cyanide	cyano
O^{2-}	oxide	oxo			

*Both are used for coordinated hydrogen, but the latter term is usually restricted to boron compounds.

The letter in each of the ligand names which is used to determine the alphabetical order is shown in bold in the following examples to illustrate the alphabetical arrangement. For many compounds, the oxidation number of the central atom and/or the charge on the ion are so well known that there is no need to use either a Stock number or a Ewens-Bassett number. However, it is not wrong to use such numbers.

$Na[B(NO_3)_4]$ sodium tetranitratoborate(1−)
                    sodium tetranitratoborate(III)

$K_2[OsCl_5N]$ potassium pentachloronitridoosmate(2−)
                 potassium pentachloronitridoosmate(VI)

$[Co(NH_2)_2(NH_3)_4]OC_2H_5$ diamidotetraamminecobalt(1+) ethoxide
                                 diamidotetraamminecobal(III) ethoxide

$Na_3[Ag(S_2O_3)_2]$ sodium bis(thiosulfate)argentite(3−)
                      sodium bis(thiosulfate)argentite(I)

$NH_4[Cr(NCS)_4(NH_3)_2]$ ammonium diamminetetrakis(isothiocyanato)chromate(1−)
                             ammonium diamminetetrakis(isothiocyanato)chromate(III)

$Ba[BrF_4]_2$ barium tetrafluorobromate(1−)
              barium tetrafluorobromate(III)

Although the common hydrocarbon radicals generally behave as anions when attached to metals, and are sometimes found to do so, their presence in coordination units is indicated by the common radical names, although they are considered anions when calculating the oxidation number.

K[B(C_6H_5)_4]	potassium tetraphenylborate(1−)
	potassium tetraphenylborate(III)
K[SbCl_5C_6H_5]	potassium pentachloro(phenyl)antimonate(1−)*
	potassium pentachloro(phenyl)antimonate(V)

*Normally, phenyl would not be placed within enclosing marks. They are used here to avoid confusion with a chlorophenyl radical.

The name of the coordinated molecule or cation should be used as it is. All neutral ligands are enclosed in brackets, with four exceptions listed below.

cis-[PtCl_2(Et_3P)_2]*	cis-dichlorobis(triethylphosphine)platinum
	cis-dichlorobis(triethylphosphine)platinum(II)
[CuCl_2(CH_3NH_2)_2]	dichlorobis(methylamine)copper
	dichlorobis(methylamine)copper(II)
[Pt(py)_4][PtCl_4]*	tetrakis(pyridine)platinum(2+)tetrachloroplatinate(2−)
	tetrakis(pyridine)platinum(II)tetrachloroplatinate(II)
[Co(en)_3]_2(SO_4)_3*	tris(ethylenediamine)cobalt(3+)sulfate
	tris(ethylenediamine)cobalt(III)sulfate
K[PtCl_3(C_2H_4)]	potassium trichloro(ethylene)platinate(1−)
	potassium trichloro(ethylene)platinate(II)
[Ru(NH_3)_5(N_2)]Cl_2	pentaamine(dinitrogen)ruthenium(2+)chloride
	pentaamine(dinitrogen)ruthenium(II)chloride

*The common abbreviations for organic groups are shown in Table 6.3.

Water and ammonia as neutral ligands in coordination complexes are called "aqua" (formerly "aquo") and "ammine", respectively. The groups NO and CO, when attached directly to a metal atom, are called "nitrosyl" and "carbonyl", respectively. When calculating the oxidation number, these ligands are treated as neutral:

[Cr(H_2O)_6]Cl_3	hexaaquachromium(3+)chloride
	hexaaquachromium(III)trichloride
Na_2[Fe(CN)_5NO]	sodium pentacyanonitrosylferrate(2−)
	sodium pentacyanonitrosylferrate(III)
K_3[Fe(CN)_5CO]	potassium carbonylpentacyanoferrate(3−)
	potassium carbonylpentacyanoferrate(II)

4.9.3 Alternative Modes of Linkage of Some Ligands

The different point of attachment of a ligand can be indicated by adding the italicized symbol(s) for the atom(s) through which attachment occurs to the end of the ligand name. For example, the dithiooxalato anion may be bound via S or O and these are distinguished as dithiooxalato-*S, S'* and dithiooxalato-*O, O '*, respectively. In some cases, other names are already in use for alternative bonding modes, such as thiocyanato(−SCN) and isothiocyanato(−NCS), nitro (−NO_2), and nitrito(−ONO).

Na$_3$[Co(NO$_2$)$_6$]	sodium hexanitrocobaltate(3−)
	sodium hexanitrocobaltate(III)
[Co(ONO)(NH$_3$)$_5$]SO$_4$	pentaamminenitritocobalt(2+) sulfate
	pentaamminenitritocobalt(III) sulfate
[Co(NCS)(NH$_3$)$_5$]Cl$_2$	pentaammineisothiocyanatocobalt(2+)chloride
	pentaammineisothiocyanatocobalt(III)chloride

The literature on coordination compounds makes extensive use of abbreviations for ligand names, especially in formulae.

4.9.4 Complexes with Unsaturated Molecules or Groups

The name of the ligand group is given with the prefix η-, which can be read as eta. If some but not all of the ligand atoms in a chain or ring, or some but not all of the ligand atoms involved in double bonds, are attached to the central atom, locant designators are inserted before η.

[Pt[Cl$_2$(NH$_3$)(C$_2$H$_4$)]	amminedichloro(η-ethylene)platinum
	amminedichloro(η-ethylene)platinum(II)
K[PtCl$_3$(C$_2$H$_4$)]	potassium trichloro(η-ethylene)platinate(1−)
	potassium trichloro(η-ethylene)platinate(II)
ReH(C$_5$H$_5$)$_2$	bis(η-cyclopentadienyl)hydridorhenium
	bis(η-cyclopentadienyl)hydridorhenium(I)

4.9.5 Compounds with Bridging Atoms or Groups

(1) A bridge group is indicated by the Greek letter μ immediately preceding its name and the name separated from the rest of the complex by hyphens.

(2) Two or more bridging groups of the same type are indicated by di-μ- or bis-μ-, etc.

(3) Bridging groups are listed with the other groups in alphabetical order, unless the symmetry of the molecule permits simpler names by the use of multiplicative prefixes.

(4) If the same ligand is present as a bridging ligand and as a non-bridging ligand, it is listed first as a bridging ligand.

[(NH$_3$)$_5$Cr-OH-Cr(NH$_3$)$_5$]Cl$_5$	μ-hydroxobis(pentaamminechromium)(5+)chloride
	μ-hydroxobis(pentaamminechromium)(III)chloride
(CO)$_3$[Fe(CO)$_3$Fe(CO)$_3$]	tri-μ-carbonylbis(tricarbonyliron)
Br$_2$Pt(SMe$_2$)$_2$PtBr$_2$	bis(μ-dimethylsulfide)bis[dibromoplatinum(II)]

4.9.6 Di- and Polynuclear Compounds without Bridging Groups; Direct Linkage between Coordination Centers

There are numerous compounds that contain metal-metal bonds. Such compounds, if symmetrical, are named using multiplicative prefixes; if unsymmetrical, one central atom and its attached ligands are treated as ligands to the other centers.

[Br$_4$Re-ReBr$_4$]$^{2-}$	bis(tetrabromorhenate)(2−) ion
	bis(tetrabromorhenate)(II) ion
[(CO)$_5$Mn-Mn(CO)$_5$]	bis(pentacarbonylmanganese)

[(CO)₄Co-Re(CO)₅]　　　　　　　　pentacarbonyl(tetracarbonylcobaltio)rhenium

[η-C₅H₅(CO)₃Mo-Mo(CO)₃-η-C₅H₅]　　bis(tricarbonyl-η-cyclopentadienyl)molybdenum

[(Cl₃Sn)₂RhCl₂Rh(SnCl₃)₂]⁴⁻　　　　di-μ-chloro-bis[(trichlorostannyl)rhodate](4−) ion

　　　　　　　　　　　　　　　　　di-μ-chloro-bis[(trichlorostannyl)rhodate](Ⅰ) ion

4.9.7 Homoatomic Aggregates

There are several instances of a finite group of metal atoms with bonds directly between the metal atoms, but also with some non-metal atoms or groups (ligands) closely associated with the cluster. The geometric shape of the cluster is denoted by triangulo, quadro, tetrahedro, octahedro, etc., and the nature of the bonds to the ligands by the conventions for bridging bonds and simple bonds. Numbers are used as locant designators, as they are for homoatomic chains and boron clusters.

Os₃(CO)₁₂　　　　　　　　dodecacarbonyl-triangulo-triosmium

Cs₃[Re₃Cl₁₂]　　　　　　　cesium dodecachloro-triangulo-trirhenate(3−)

　　　　　　　　　　　　　cesium dodecachloro-triangulo-trirhenate(Ⅲ)

B₄Cl₄　　　　　　　　　　tetrachloro-tetrahedro-tetraboron

[Nb₆Cl₁₂]²⁺　　　　　　　 dodeca-μ-chloro-octahedro-hexaniobium(2+) ion

4.10 Addition Compounds

This rule covers some donor-acceptor complexes and a variety of lattice compounds. It is particularly relevant for compounds of uncertain structure; new structural information often allows naming according to the rules for coordination compounds.

The suffix -ate is now the accepted suffix for anions and should generally not be used for addition compounds. Alcoholates are the salts of alcohol and this name should not be used to indicate the alcohol of crystallization. Similarly, addition compounds containing ether, ammonia, etc. should not be called etherates, ammoniates, etc.

There is, however, one exception. According to the generally accepted meaning of the suffix -ate, "hydrate" would be, and has been, the name for a salt of water, i.e., what is now known as a hydroxide. The name hydrate is now very well established as the name of a water-containing or crystallizing compound and is also permitted in these rules to designate water bound in some unspecified way. Even in this case, it is considered preferable to avoid the suffix -ate by using the name "water" or its equivalent in other languages whenever possible.

The names of addition compounds may be formed by connecting the names of the individual compounds by hyphens and indicating the number of molecules after the name by Arabic numerals separated by the solidus. Boron compounds and water are always listed last in this order. Other molecules are listed in order of increasing number, and those occurring in equal numbers are listed in alphabetical order.

Na₂CO₃·10H₂O　　　　　sodium carbonate decahydrate

　　　　　　　　　　　　sodium carbonate 10-water

　　　　　　　　　　　　sodium carbonate-water(1/10)

KAl(SO₄)₂·12H₂O　　　　potassium aluminum sulfate dodecahydrate

　　　　　　　　　　　　aluminum sulfate-potassium sulfate-water(1/1/24)

CaCl$_2$·8NH$_3$	calcium chloride-ammonia(1/8)
AlCl$_3$·4EtOH	aluminum chloride-ethanol(1/4)
BF$_3$·2MeOH	methanol-boron trifluoride(2/1) (esterification reagent)
BF$_3$·2NH$_3$	ammonia-boron trifluoride(2/1)

4.11 Prefixes or Affixes Used in Inorganic Nomenclature

Multiplying affixes include (1) mono, di, tri, tetra, penta, hexa, hepta, octa, nona (ennea), deca, undeca (hendeca), dodeca, etc., used by direct joining without hyphens; and (2) bis, tris, tetrakis, pentakis, etc., used by direct joining without hyphens but usually with enclosing marks around each whole expression to which the prefix applies. Structural affixes are usually italicized and separated from the rest of the name by hyphens.

Note the difference between the use of this prefix in inorganic and organic chemistry. In organic chemistry, *meta* indicates the positions of substituents in aromatic cyclic compounds. The substituents are at the 1,3-positions, for example in resorcinol. *Ortho* describes a molecule with substituents at the 1 and 2 positions in an aromatic compound. In other words, the substituent is next to or adjacent to the primary carbon on the ring and the symbol for *ortho* is o- or 1,2-. In inorganic chemistry, however, ortho and meta have a different meaning when describing inorganic acids than the same prefixes have when describing benzene derivatives. Orthophosphoric acid can be defined as a weak mineral acid and has as much water in its formula as possible with the chemical formula H$_3$PO$_4$. Pyro- and metaphosphoric acids are inorganic compounds with the chemical formulae H$_4$P$_2$O$_7$ and HPO$_3$ respectively. Dehydration of orthophosphate forms metaphosphoric acid.

Some of the geometrical and structural prefixes used in the inorganic nomenclature suggested by IUPAC are shown in Table 4.1.

Table 4.1 Some geometrical and structural affixes used in inorganic nomenclature

Affixes	Meaning
antiprismo	an affix used in names to denote eight atoms bound into a rectangular **antiprism**
asym	asymmetrical
catena	a chain structure; often used to designate linear polymeric substances
closo	a cage or closed structure, especially a boron skeleton that is a polyhedron having all faces triangular
cyclo	a ring structure. (here, cyclo is used as a modifier indicating structure and hence is italicized. in organic nomenclature, "cyclo" is considered to be part of the parent name since it changes the molecular formula. it is therefore not italicized).
dodecahedro	eight atoms bound into a dodecahedron with triangular faces. (A dodecahedron is a polyhedron with 12 faces. 2 different types of dodecahedrons are in the figure below with the prefix "dodeca" meaning twelve.)

(continued)

Affixes	Meaning
fac	three groups occupying the corners of the same face of an octahedron
hexahedro	eight atoms bound into a hexahedron (e.g. cube)
hexaprismo	twelve atoms bound into a hexagonal prism
icosahedro	twelve atoms bound into a triangular icosahedron
mer	meridional; three groups occupying vertices of an octahedral coordination sphere in such a relationship that one is *cis* to the two others which are themselves *trans*
nido	a nest-like structure, especially a boron skeleton that is almost closed
octahedro	six atoms bound into an octahedron
pentaprismo	ten atoms bound into a pentagonal prism
quadro	four atoms bound into a quadrangle (e.g. square)
sym	symmetrical
tetrahedro	four atoms bound into a tetrahedron
cis[①]	two groups occupying adjacent positions in a coordination sphere
trans[①]	two groups occupying positions in a coordination sphere directly opposite each other, i.e. in the polar positions of a sphere
triangulo	three atoms bound into a triangle
triprismo	six atoms bound into a triangular prism
η (eta)	specifies the bonding of contiguous atoms of a ligand to a central atom
μ (mu)	signifies that a group so designated bridges two or more coordination centers
σ (sigma)	signifies that one atom of the group is attached to a metal

① In chemistry, *cis*- and *trans*- are descriptors that indicate the relationship between two ligands that are attached to separate atoms that are connected by a double bond or are contained in a ring. The two ligands are said to be *cis* to each other if they are on the same side of a plane. If they are on opposite sides, their relative position is called *trans*. The appropriate reference plane of a double bond is perpendicular to that of the relevant σ bonds and passes through the double bond. For a ring (the ring being in a conformation, real or assumed, without re-entrant angles at the two substituted atoms) it is the mean plane of the ring(s). For alkenes, the terms *cis* and *trans* can be ambiguous and have therefore been largely replaced by the *E*, *Z* convention for the nomenclature of organic compounds. When there are more than two entities attached to the ring, the use of *cis* and *trans* requires the definition of a reference substituent.

New Words and Expressions

nomenclature	[nəˈmeŋklətʃə(r)]	n. 命名法，术语
inorganic	[ˌɪnɔːˈɡænɪk]	adj. 无机的，无生物的
proliferation	[prəˌlɪfəˈreɪʃn]	n. 激增，剧增，繁殖，增生，大量
non-systematic	[ˈnɒnsɪstəmˈætɪk]	adj. 非系统化，非系统性的
ethanoate	[eˈθeɪˌnəʊt]	n. 醋酸盐，醋酸酯
acetate	[ˈæsɪteɪt]	n. 醋酸盐
acquaint	[əˈkweɪnt]	v. 使熟悉，使了解
coordination	[kəʊˌɔːdɪˈneɪʃ(ə)n]	n. 协调，配合，配位
crystallographer	[ˌkrɪstəˈlɒɡrəfə(r)]	n. 检晶器
multiply	[ˈmʌltɪplaɪ]	v. 增加，繁殖，乘
affix	[əˈfɪks]	n. 词缀
stoichiometry	[ˌstɔɪkɪˈɒmɪtri]	n. 化学计量学

tetraamminedichlorocobalt		n. 四氨基二氯钴
omit	[əˈmɪt]	v. 遗漏，省略
diphosphoric acid		焦磷酸
disilane	[daɪˈsɪleɪn]	n. 乙硅烷
tetrathionate	[tetrəˈθaɪəneɪt]	n. 连四硫酸盐
ambiguity	[ˌæmbɪˈgjuːəti]	n. 模棱两可，不明确，含混不清的语句
ligand	[ˈlɪgənd]	n.（生化）配位体，配基
chloroethyl	[kləˈrəʊɪθɪl]	n. 氯乙基
decyl	[diːˈsɪl]	adj. 癸基的，烷基的；n. 癸基，烷基
phosphine	[ˈfɒsfiːn]	n. 磷化氢，三氧化磷
hydroxymethyl	[haɪdrɒksɪˈmeθɪl]	n. 羟甲基
phosphonium	[fɒsˈfəʊnɪəm]	n. 磷（根），磷（一价基）
chloride	[ˈklɔːraɪd]	n. 氯化物
methyl	[ˈmeθɪl]	n. 甲基，木精
isocyanide	[aɪsəʊˈsaɪənaɪd]	n. 异腈，异氰化物
parentheses	[pəˈrenθəsiːz]	n. 圆括号，插入成分（parenthesis的复数）
substituent	[sʌbˈstɪtjuənt]	adj. 被代替的；n. 取代基，取代者
chlorotetrasilane		n. 氯四硅烷
aluminum	[əˈluːmɪnəm]	n. 铝
potassium	[pəˈtæsiəm]	n. 钾
decahydrate	[ˌdekəˈhaɪdreɪt]	n. 十水合物
hexaborane	[ˈheksəbərən]	n. 六硼烷
hexaboron		n. 六硼
italic	[ɪˈtælɪk]	adj. 斜体的；n. 斜体字
orthographic	[ˌɔːθəˈgræfɪk]	adj. 正字法的，拼字正确的，正交的
o-methylhydroxylamine		n. 邻甲基羟胺
cysteinato		n. 半胱氨酸负离子
abbreviation	[əˌbriːviˈeɪʃn]	n. 缩略词，缩写形式，缩略，缩写
capitalize	[ˈkæpɪtəlaɪz]	v. 用大写字母书写（或印刷），首字大写，资本化
fusion	[ˈfjuːʒ(ə)n]	n. 融合，核聚变，熔化，融合物
fission	[ˈfɪʃ(ə)n]	n. 裂变，分裂，分体，分裂生殖法
decay	[dɪˈkeɪ]	v. 腐朽，腐烂；n. 腐烂
sequence	[ˈsiːkwəns]	n. 顺序，次序；v. 按顺序排列
heteropolyatomic		adj. 杂多原子的
calcium	[ˈkælsiəm]	n. 钙
hydrogen	[ˈhaɪdrədʒən]	n. 氢，氢气

peroxide	[pəˈrɒksaɪd]	n. 过氧化氢，过氧化物
dichloride	[daɪˈklɔːraɪd]	n. 二氯化合物
gallium	[ˈɡæliəm]	n. 镓
arsenic	[ˈɑːsnɪk]	n. 砷，砒霜；adj. 砷的，五价砷的
lithium	[ˈlɪθiəm]	n. 锂
nickel	[ˈnɪk(ə)l]	n. 镍；v. 镀镍于
silicon	[ˈsɪlɪkən]	n. 硅
difluoride	[daɪfˈluəraɪd]	n. 二氟化物
carbon	[ˈkɑːbən]	n. 碳；adj. 碳的，碳处理的
sulfur	[ˈsʌlfə(r)]	n. 硫
polyatomic	[ˌpɒliəˈtɒmɪk]	adj. 多原子的，多碱的，多酸的
binary	[ˈbaɪnəri]	adj. 由两部分组成的，二重的；n. 二进制
nitrogen	[ˈnaɪtrədʒən]	n. 氮
bismuthine	[ˈbɪzməθɪn]	n. 辉铋矿，氢化铋
diborane	[daɪˈbɒreɪn]	n. 乙硼烷
plumbane	[pˈlʌmbeɪn]	n. 铅烷
trisilane	[ˈtraɪsɪleɪn]	n. 丙硅烷
polysulfane	[pɒliːsʌlˈfæn]	n. 聚硫烷
modification	[ˌmɒdɪfɪˈkeɪʃ(ə)n]	n. 修改的行为（过程），修改，更改，修饰
copper	[ˈkɒpə(r)]	n. 铜；adj. 紫铜色的；v. 给……镀铜
cuprous	[ˈkjuːprəs]	adj. 亚铜的，一价铜的
cupric	[ˈkjuːprɪk]	adj. 铜的，含铜的，二价铜的
ferrous	[ˈferəs]	adj. 亚铁的，铁的，含铁的
ferric	[ˈferɪk]	adj. 铁的，三价铁的，含铁的
mercury	[ˈmɜːkjəri]	n. 汞，水银，水银柱（Mercury 水星）
mercurous	[ˈmɜːkjurəs]	adj. 一价汞的，（亚）汞的，含汞的
mercuric	[mɜːˈkjuərɪk]	adj. 汞的，水银的
lead	[liːd]	v. 导致，造成；n. 铅
plumbous	[ˈplʌmbəs]	adj. 铅的
plumbic	[ˈplʌmbɪk]	adj. 铅的，含四价铅的
tin	[tɪn]	n. 锡
stannous	[ˈstænəs]	adj. 锡的，含锡的，含二价锡的
stannic	[ˈstænɪk]	adj. 锡的，四价锡的，含锡的
iodine	[ˈaɪədiːn]	n. 碘，碘酒
nitrosyl	[ˈnaɪtrəsɪl]	n. 亚硝酰基；adj. 亚硝酰基的
nitryl	[ˈnaɪtrɪl]	n. [化学]硝基，硝基二甲苯
permanganate	[pəˈmæŋɡənet]	n. 高锰酸盐

thiocyanate	[ˌθaɪəʊˈsaɪəˌneɪt]	n. 硫氰酸盐
hydrazinium	[haɪdrəˈzɪnɪəm]	n. 肼
anilinium		n. 苯胺离子，苯胺基
glycinium		n. 甘氨酸阳离子
pyridinium	[pɪrɪˈdɪnɪəm]	n. 吡啶
abbreviate	[əˈbriːvɪeɪt]	v. 缩写，省略
azide	[ˈeɪzaɪd]	n. 叠氮化物
triiodide	[traɪˈaɪəʊdaɪd]	n. 三碘化物
peroxynitrite	[pərɒksɪnɪtˈraɪt]	n. 过氧硝酸盐，过氧亚硝基
disulfite		n. 二亚硫酸盐
chalcogen	[ˈkælkədʒən]	n. 硫族元素，氧族元素
irrespective	[ˌɪrɪˈspektɪv]	adj. 不考虑，不顾
uranyl	[ˈjʊərənɪl]	n. 铀酰
neptunyl	[nepˈtjuːnɪl]	n. 镎酰
sulfinyl	[ˈsʌfɪnɪl]	n. 亚磺酰基（同 sulphinyl）
sulfonyl	[ˈsʌfənɪl]	n. 磺酰，磺酰基（同 sulphonyl）
fluoro	[ˈfluərə]	n. 氟代
chlorosulphonic acid		氯磺酸
fluorosulfonic acid		氟磺酸
chlorosulfuric acid		氯磺酸，硫酰氯
amidosulfuric acid		氨基磺酸
oxoacid	[ˌɒksəʊˈæsɪd]	n. 酮酸
hypochlorite	[ˌhaɪpəʊˈklɔːraɪt]	n. 次氯酸盐，低氧化氯
hypochlorous	[ˌhaɪpəʊˈklɔːrəs]	adj. 次氯酸的
chlorite	[ˈklɔːraɪt]	n. 绿泥石，亚氯酸盐
chlorous	[ˈklɔːrəs]	adj. 与氯化合的，亚氯酸的
chlorate	[ˈklɔːreɪt]	n. 氯酸盐
chloric	[ˈklɔːrɪk]	adj. 氯的，含氯的
perchlorate	[pəˈklɔːreɪt]	n. 高氯酸盐（或酯）
perchloric	[pəˈklɔːrɪk]	adj.（含）高氯的
orthoboric	[ˌɔːθəʊˈbɔrɪk]	adj. 正硼酸的
metaboric acid		偏硼酸
orthosilicic acid		原硅酸
metasilicic acid		正硅酸
orthophosphoric acid		正磷酸
metaphosphoric acid		偏磷酸
orthoperiodic acid		原高碘酸
orthotelluric acid		原碲酸
indicate	[ˈɪndɪkeɪt]	v. 表明，要求，写明，简要陈述
irregular	[ɪˈregjələ(r)]	adj. 不平整的，不整齐的
hydrogencarbonate	[ˈhaɪdrədʒenˈkɑːbənɪt]	n. 碳酸氢盐

bicarbonate	[ˌbaɪˈkɑːbənet]	n. 碳酸氢盐，重碳酸盐，酸式碳酸盐
dihydrogenphosphate		n. 磷酸二氢盐
hydrogenphosphate		n. 磷酸氢盐
hydrogensulfide		n. 硫化氢
hydrogensulfite		n. 酸式亚硫酸盐，亚硫酸氢根离子
bisulfite	[ˌbaɪˈsʌlfaɪt]	n. 重亚硫酸盐，酸性亚硫酸盐
pentahydrate	[pentəˈhaɪdreɪt]	n. 五水合物
chromium	[ˈkrəumiəm]	n. 铬
nitrate	[ˈnaɪtreɪt]	n. 硝酸盐；v. 用硝酸处理
nonahydrate		n. 九水合物
dihydrate	[daɪˈhaɪdreɪt]	n. 二水合物
monohydrate	[ˌmɒnəuˈhaɪdreɪt]	n. 一水合物，一水化物
potassium	[pəˈtæsiəm]	n. 钾
dodecahydrate	[ˌdəudɪkəˈhaɪdreɪt]	n. 十二水合物
tripotassium	[trɪpəuˈtæsiəm]	n. 三钾
tetrapotassium		n. 四钾
dimethylamine	[daɪmeθɪləˈmiːn]	n. 二甲胺
dichlorodiammine		n. 二氯二胺
diamminedichloroplatinum	[diːæmɪndɪkˈlɒrəuplætmən]	n. 顺铂，顺二氯化二氨亚铂
triphosphato		n. 三磷酸
thiosulphate	[ˌθaɪə(u)ˈsʌlfeɪt]	n. 硫代硫酸盐
dimethylamido		n. 二甲酰胺
alphabetical	[ˌælfəˈbetɪkl]	adj. 按字母顺序的，字母（表）的
pentachloronitridoosmate		n. 五氯氮化合物
isothiocyanato	[aɪsəθiːəusaɪəˈneɪtəu]	n. 异硫氰基，异硫氰酸酯
tetrafluorobromate		n. 四氟溴酸盐
tetraphenylborate	[tetˈræfənɪlbəreɪt]	n. 四苯基硼酸，四苯硼酸根
ethylenediamine	[ˈeθəliːnˈdaɪəmiːn]	n. 乙二胺
calculating	[ˈkælkjuleɪtɪŋ]	adj. 精明的，精于算计的
pentacyanonitrosylferrate		n. 五氰基亚硝基高铁酸盐
carbonnylpentacyanoferrate		n. 羰基五氰基高铁酸盐
dithiooxalato		n. 二硫代草酸酯
hexanitrocobaltate		n. 六硝基钴酸盐
pentaamminenitritocobalt		n. 五胺腈
pentaammineisothiocyanatocobalt		n. 五胺邻硫氰酸钴
unsaturated	[ʌnˈsætʃəˌreɪtɪd]	adj. 不饱和的
amminedichloro		n. 氨基二氯
dibromoplatinum		adj. 二溴铂
polynuclear	[ˌpɒlɪˈnjuːklɪə]	adj. 多核的
unsymmetrical	[ˌʌnsɪˈmetrɪkəl]	adj. 非对称的，不匀称的
tetrabromorhenate		n. 四溴铼酸盐

pentacarbonylmanganese		n.	五羰基锰
molybdenum	[məˈlɪbdənəm]	n.	钼
homoatomic	[ˌhəʊməʊˈtɒmɪk]	adj.	同原子的，同素（种）的
aggregate	[ˈæɡrɪɡət]	adj.	总计的，合计的；v. 集合，聚集
tetrahedron	[ˌtetrəˈhiːdrən]	n.	四面体
octahedron	[ˌɒktəˈhiːdrən]	n.	八面体
dodecacarbonyl		n.	十二羰基
cesium	[ˈsiːzɪəm]	n.	铯
dodecachloro		n.	十二氯
tetrachloro	[tetˈrəklərəʊ]	n.	四氯
crystallization	[ˌkrɪstəlaɪˈzeɪʃ(ə)n]	n.	结晶，晶化，结晶体
unspecified	[ʌnˈspesɪfaɪd]	adj.	未指明的，未详细说明的
esterification	[eˌsterɪfɪˈkeɪʃən]	n.	酯化
benzene	[ˈbenziːn]	n.	苯
antiprism	[ˈæntiˌprɪzəm]	n.	反棱柱，反棱镜
asymmetrical	[ˌeɪsɪˈmetrɪk(ə)l]	adj.	非对称的，不匀称的，不对等的
chain	[tʃeɪn]	n.	链，链条；v. 用锁链拴住，拘禁，束缚
polyhedron	[ˌpɒliˈhiːdrən]	n.	多面体
hexahedron	[ˌheksəˈhiːdrən]	n.	六面体
hexagonal	[hekˈsæɡən(ə)l]	adj.	六边的，六角形的
prism	[ˈprɪzəm]	n.	棱镜，棱柱
meridional	[məˈrɪdɪənl]	adj.	子午线的，南部的；n. 南欧人
occupy	[ˈɑːkjupaɪ]	v.	占用，占领
vertices	[ˈvɜːtɪsiːz]	n.	顶点，天顶，头顶（vertex 的复数）
skeleton	[ˈskelɪt(ə)n]	n.	骨骼，骨架，框架
pentagonal	[penˈtæɡənl]	adj.	五角的，五边形的
adjacent	[əˈdʒeɪs(ə)nt]	adj.	邻近的，毗连的，相邻的

Notes

1. The oxidation number of an element in a chemical compound is the charge that would be present on an atom of the element if the electrons in each bond were assigned to the more electronegative atom.
   化合物中元素的氧化数是指如果每个键中的电子被分配到电负性更强的原子上，该元素原子上的电荷数。
2. The multiplicative prefixes bis, tris, tetrakis, pentakis, etc. were originally introduced into organic nomenclature to denote a set of identical radicals, each substituted in the same way, and are sometimes used to avoid ambiguity.

倍数前缀 bis, tris, tetrakis, pentakis 等最初被引入有机物命名法，表示一组相同的自由基，每个自由基以相同的方式替换，有时用于避免歧义。

3. Since the charge of the coordination entity is the algebraic sum of the charges of its constituents, the necessary information may be provided by giving either the Stock number (formal charge of the central ion, i.e., oxidation number) or the Ewens-Bassett number.

   由于配位单元的电荷数是其组成部分电荷的代数和，因此可以通过给出 Stock 数（中心离子的形式电荷，即氧化数）或 Ewens-Bassett 数来提供必要的信息。[In English oxidation number is largely synonymous with oxidation state, and may be preferred when the value represents a mere parameter or number rather than being related to chemical systematics or a state of the atom in a compound. Etymologically, it stems from the no-longer-used term Stock number (oxidation number of a central atom; the charge it would bear if all the ligands were removed along with the electron pairs that were shared with the central atom) and the likewise obsolete term Ewens-Bassett number (ion charge). Stock 数也称为"库存数" Ewens-Bassett 数指的是离子电荷数，这些都是比较早期的对于氧化数的描述，越来越多的教材已经用 oxidation number 氧化数来替代；https://goldbook.iupac.org/terms/view/O04363]

Trivial — Last but not the Least

Table 4.2　Multiplicative prefixes for simple and complicated entities

No.	Simple	Complicated	No.	Simple	Complicated
2	di	bis	8	octa	octakis
3	tri	tris	9	nona	nonakis
4	tetra	tetrakis	10	deca	decakis
5	penta	pentakis	11	undeca	undecakis
6	hexa	hexakis	12	dodeca	dodecakis
7	hepta	heptakis	20	icosa	icosakis

Unit 5
A Short Introduction to Organic Chemistry

5.1　Overview

The word organic literally means "derived from living organisms". Originally, the science of organic chemistry was the study of compounds such as sugar, urea, starch, waxes, and vegetable oils, which were considered "organic" and people accepted vitalism: the belief that natural products require a "vital force" to create them. Organic chemistry is more than just the study of carbon or the study of chemicals in living organisms. Organic chemistry is everywhere and is therefore the study of compounds with "vital force". Inorganic chemistry was the study of gases, rocks, minerals, and the compounds that could be made from them. Organic chemistry is the study of carbon and the study of the chemistry of life. Since not all carbon reactions are organic, another way of looking at organic chemistry would be to consider it the study of molecules containing the carbon-hydrogen (C—H) bond and their reactions. Organic chemistry is important because it is the study of life and all the chemical reactions associated with life. Several professions use an understanding of organic chemistry, such as doctors, veterinarians, dentists, pharmacologists, chemical engineers, and chemists. Organic chemistry plays a role in the development of common household chemicals, foods, plastics, medicines, and fuels — most of the chemicals that are part of everyday life.

In the 19th century, experiments showed that organic compounds could be synthesized from inorganic compounds. In 1828, the German chemist Friedrich Wöhler converted ammonium cyanate, made from ammonia and cyanic acid, into urea simply by heating it in the absence of oxygen. Urea has always come from living organisms and was thought to contain the vital force, but ammonium cyanate is inorganic and therefore lacks the vital force. Some chemists claimed that a trace of vital force from Wöhler's hand must have contaminated the reaction, but most recognized the possibility of synthesizing organic compounds from inorganic ones. Many other syntheses were carried out and the "vital force" theory was eventually discarded.

Since vitalism was disproved in the early nineteenth century, you might think it would have died out by now. You would be wrong! Vitalism lives on today in the minds of those who believe that "natural" (herbal) vitamins, flavorings, etc. are somehow different and healthier than identical "artificial" (synthetic) compounds. As chemists, we know that herbal and synthetic compounds are identical. Assuming they are pure, the only way to tell them apart is by ^{14}C dating. Compounds synthesized from petrochemicals have lower levels of radioactive ^{14}C and appear old because their ^{14}C has decayed over time. More recently, compounds from plants have been synthesized from CO_2 in the air. These have high levels of radioactive ^{14}C. Some major chemical suppliers offer isotope ratio analysis to show that "natural" has a high ^{14}C content and is of plant origin. Such sophisticated analysis gives a high-tech flavor to this 21st century form of vitalism.

Although organic compounds do not require a vital force, they are still different from inorganic compounds. The characteristic feature of organic compounds is that they all contain one or more carbon atoms. But not all carbon compounds are organic substances. Diamond, graphite, carbon dioxide, ammonium cyanate and sodium carbonate are derived from minerals and have typical inorganic properties, but most of the millions of carbon compounds are classified as organic.

The modern definition of organic chemistry is the chemistry of carbon compounds. What is so special about carbon that an entire branch of chemistry is devoted to its compounds? Unlike most other elements, carbon forms strong bonds with other carbon atoms and with a wide range of other elements. Chains and rings of carbon atoms can form an infinite variety of molecules. It is this diversity of carbon compounds that provides the basis for life on Earth. Living things are largely made up of complex organic compounds that serve structural, chemical, or genetic functions.

Life is based on the ability of carbon to combine with other carbon atoms to form diverse structures. Reflecting this fact, the branch of chemistry that studies carbon-containing compounds is known as organic chemistry. Today, more than 13 million organic compounds are known, and about 100 000 new ones are discovered every year. These include those discovered in nature and those synthesized in the laboratory (in contrast, only 200 000 to 300 000 inorganic compounds based on elements other than carbon are known).

Although carbon is not one of the ten most abundant elements in the Earth's crust, surface waters, and atmosphere, it is the third most abundant element in the human body in terms of number of atoms and the second most abundant in terms of mass[1]. We ourselves are largely made up of organic molecules, and we are nourished by the organic compounds in our food. The proteins in our skin, the lipids in our cell membranes, the glycogen in our livers and the DNA in the nuclei of our cells are all organic compounds. Our bodies are also regulated and defended by complex organic compounds. Water makes up 60% - 90% of all living matter. However, of the remaining 10% - 40% of living matter, less than 4% is inorganic, the rest is organic.

Chemists have learned to synthesize or simulate many of these complex molecules. The synthetic products serve as drugs, medicines, plastics, pesticides, paints, and fibers. Many of the most important advances in medicine are, in fact, advances in organic chemistry. New synthetic drugs are developed to fight disease, and new polymers are molded to replace failing organs. Organic chemistry has come full circle. It began as the study of compounds derived from "organs", and now it gives us the drugs and materials we need to save or replace those organs.

Organic molecules are classified according to the functional groups they contain, where a functional group is defined as a combination of atoms that behave as a unit. Most functional groups are distinguished by the heteroatoms they contain. Carbon atoms bond to each other and to hydrogen atoms in many different ways, resulting in an incredibly large number of hydrocarbons. But carbon atoms can also bond to atoms of other elements, further increasing the number of possible organic molecules. In organic chemistry, any atom other than carbon and hydrogen in an organic molecule is called a heteroatom, where hetero means "other than C or H". The role that heteroatoms play in determining the proper bonding of each class is the underlying theme. As you study the section on heteroatoms, focus on understanding the chemical and physical properties of the different classes of compounds, as this will give you a greater appreciation for the remarkable diversity of organic molecules and their many applications.

A hydrocarbon structure can act as a framework to which various heteroatoms can be attached. This is analogous to a Christmas tree acting as a framework on which ornaments are hung. Just as the ornaments give character to the tree, heteroatoms give character to an organic molecule. In other words, heteroatoms can have a profound effect on the properties of an organic molecule. Consider ethane, C_2H_6, and ethanol, C_2H_6O, which differ only by a single oxygen atom. Ethane has a b.p. of −88℃, making it a gas at room temperature, and it does not dissolve very well in water. Ethanol, on the other hand, has a b.p. of +78℃, making it a liquid at room temperature. It is infinitely soluble in water and is the active ingredient in alcoholic beverages. Consider also ethylamine, C_2H_7N, which has one nitrogen atom and is a corrosive, pungent, highly toxic gas, very different from ethane or ethanol.

Organic chemistry is important because it is the study of life and all the chemical reactions associated with life. Organic chemistry can be divided into three interrelated areas: structure determination, reactions and synthesis, and the mechanisms by which the reactions take place. Organic chemistry is too large to be summarized briefly, and an accurate perception of modern organic chemistry can only be gained through examining typical examples of reactions, syntheses and mechanisms.

5.2 Types of Organic Chemical Reactions

The breakdown process of substances into their constituent atoms is common for all compounds, whether organic or inorganic. Additionally, all organic compounds undergo a reaction with oxygen that results in combustion and the production of constituent atom oxides.

The importance of the combustion reaction cannot be minimized: most of the heat and energy (apart from that generated by falling water, wind, sun, compressed steam, and within the Earth or by nuclear sources) comes from the combustion of very complicated oxidative processes in which carbon compounds are ultimately converted to carbon dioxide and water[2]. The combustion reactions are usually not very interesting. They are rarely captured by reactions that change only a few bonds in the molecule at a time. Selective reactions are those in which relatively small changes occur in the conversion of reactants to products. Hundreds of such reactions are now known to organic chemists, many of which are introduced in the specialized field of organic synthesis.

To introduce this large and important segment of organic chemistry, we will first focus on some of the reactions that are characteristic of the different types of carbon scaffolds that we have discussed. To do this, we will organize the discussion around four main types of reactions: **addition reactions, substitution reactions, elimination reactions, and oxidation reactions**.

Addition reactions of alkenes. The most generally useful reactions of alkenes involve the addition of various reagents to the double bond, converting the double bond to a single bond in the processes shown in Figure 5.1.

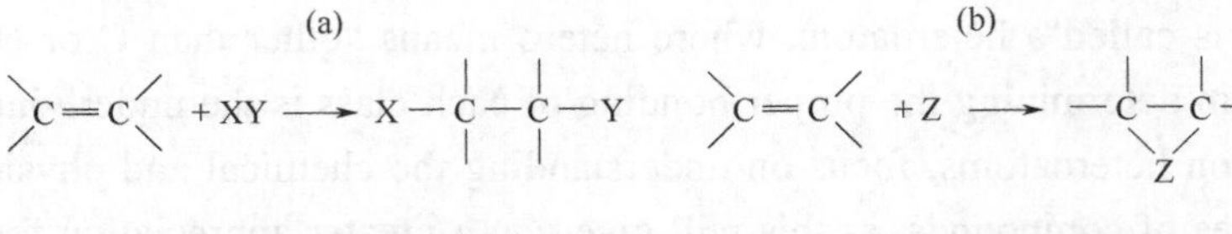

(a) $\text{C}=\text{C} + XY \longrightarrow X-\text{C}-\text{C}-Y$

(b) $\text{C}=\text{C} + Z \longrightarrow -\text{C}-\text{C}-$ (with Z)

Figure 5.1 Diagram of the addition reactions of alkenes

Although the majority of addition processes involving alkenes occur via carbonium ion intermediates, a few follow a different route and involve radical intermediates. One of the most interesting of these is the addition of hydrogen bromide in the presence of peroxide reagents. In the absence of peroxide, hydrogen bromide adds to propene via the Markovnikov pathway to give isopropyl bromide. In the presence of peroxides, however, the order of addition is reversed and the product is *n*-propyl bromide; the addition in this case is said to be anti-Markovnikov. This is interpreted to mean that the addition reaction is initiated by the bromine atom, Br · , rather than by a proton as in the electrophilic addition. The bromine atom formed by the action of a free radical on hydrogen bromide (e.g., RO · + HBr $\longrightarrow$ ROH + Br ·) adds to the double bond of propene either at C-1 to form a secondary radical or at C-2 to form a primary radical. Since the stabilities of the radicals follow the same order as that of the carbonium ions, the reaction proceeds preferentially via the secondary radical, which then collides with a molecule of hydrogen bromide to form *n*-propyl bromide and a bromine atom. Once this stage of the reaction sequence is reached, the process becomes self-perpetuating; each time a carbon radical extracts a hydrogen from hydrogen bromide, another bromine atom is formed, which adds itself to propene to form another carbon radical[3]. If an unsaturated bond undergoes an addition reaction with a hydrogen molecule, this may also be designated as a **reduction reaction**.

Substitution reactions of alkanes. The most characteristic reaction of alkanes, which are much less reactive compounds than alkenes, alkynes or arenes, is a substitution reaction in which a hydrogen is replaced by another atom or group, e.g., R–H + Y · $\longrightarrow$ R–Y + H · . Of the many examples of reactions of this type, the most useful generally involve chlorine or bromine. For example, when a mixture of *n*-butane and chlorine is heated or irradiated, a reaction takes place to produce a mixture of 1-chlorobutane and 2-chlorobutane, as depicted in Figure 5.2.

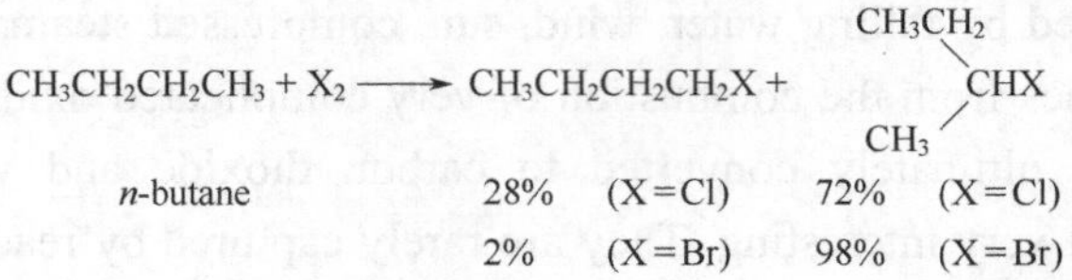

$CH_3CH_2CH_2CH_3 + X_2 \longrightarrow CH_3CH_2CH_2CH_2X + \begin{matrix} CH_3CH_2 \\ CHX \\ CH_3 \end{matrix}$

n-butane   28% (X=Cl)   72% (X=Cl)
              2% (X=Br)   98% (X=Br)

Figure 5.2 Diagram of the substitution reactions of alkanes

There is a great deal of evidence that this reaction takes place with the dissociation of molecular chlorine into chlorine atoms ($Cl_2 = Cl · + Cl ·$), and that a chlorine atom then collides with a molecule of *n*-butane to extract a hydrogen atom, forming hydrogen chloride and a butyl radical. If the collision occurs on the C—H bond of the methyl group, a primary radical is formed ($CH_3CH_2CH_2\dot{C}H_2$); if it occurs on the C—H bond of the methylene carbon, a secondary radical is formed ($CH_2CH_2\dot{C}HCH_3$). Subsequent collisions of these radicals with molecular chlorine then give 1-chlorobutane and 2-chlorobutane, respectively, together with chlorine atoms. At this stage of the reaction sequence, we have the same situation as in the peroxide-induced addition of hydrogen bromide to alkenes, namely, a self-perpetuating chain reaction. As in that case, the chain does not continue indefinitely, as various chain terminating reactions occur, requiring a constant supply of new chlorine atoms to keep the reaction going. However, up to 10^4 cycles of self-perpetuating steps can take place before the chain is broken.

Elimination reactions. Addition reactions, as we have introduced previously, proceed from a less saturated reactant to a more saturated product. Elimination reactions are the reverse case and proceed from a more saturated reactant to a less saturated product. An elimination reaction is a reaction in which atoms are removed as molecules or compounds. Elimination is generally catalyzed by a metal, an acid or base. A comparable reaction, more generally useful for laboratory synthesis, is the partial removal of hydrogen from six-membered alicyclic compounds to give aromatic compounds. For example, when cyclohexane is treated with platinum, palladium, sulfur or selenium at elevated temperatures, hydrogen is removed to form benzene, as shown in Figure 5.3. Similarly, tetralin and decalin can be converted to naphthalene:

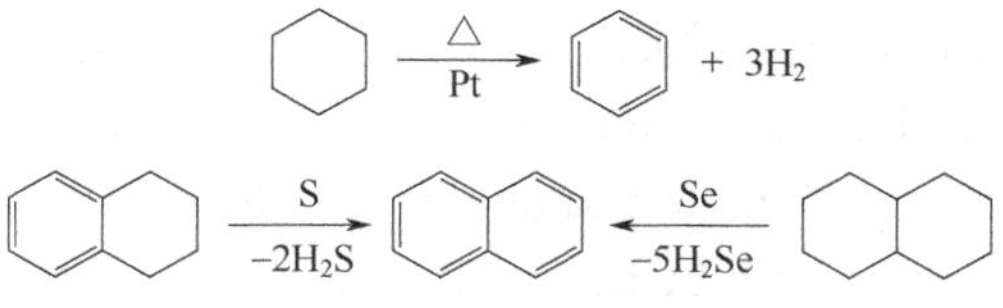

Figure 5.3 Dehydrogenation of benzene, tetralin, and decalin

Oxidation reactions of alkenes and Alkynes. For the purposes of this discussion, oxidation reactions are defined as processes in which C—H or C—C bonds are converted to C—O bonds, or in which carbon-carbon multiple bonds are converted to C=O bonds.

Alkenes and alkynes are much less resistant to oxidation than alkanes and undergo cleavage of the C=C, or C≡C bond in the presence of a variety of oxidizing agents. A particularly useful oxidizing agent is ozone, which converts alkenes of the structure RCH=CH_2 or RCH=CHR' to aldehydes, alkenes of the structure R_2C=CR'_2 to ketones and alkenes of the structure R_2C=CHR' to mixtures of an aldehyde and a ketone. Alkynes can also be converted to carboxylic acids with ozone, but the reaction is slower than with alkenes and may yield mixtures containing other compounds in addition to or instead of the carboxylic acid. Ozonolysis is a very useful way to find the position of a C=C bond in a compound. For example, suppose we have a sample of a material that we know is either compound (a) or compound (b) in Figure 5.4. If it is compound (a), then both products of ozonolysis are ketones; if it is compound (b), then one of the products of ozonolysis is a ketone and the other is an aldehyde.

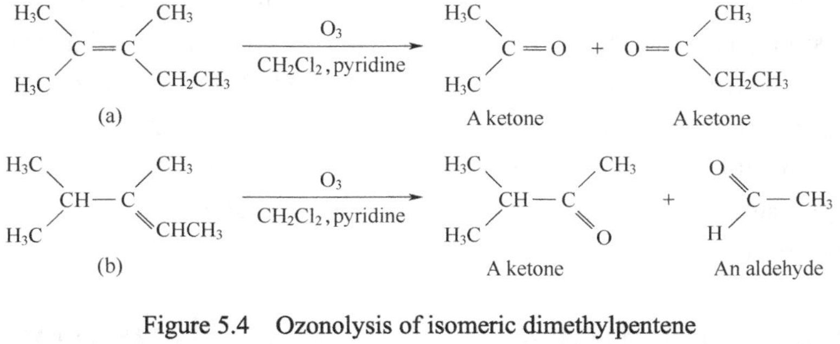

Figure 5.4 Ozonolysis of isomeric dimethylpentene

Of course, in order to be able to use reactions such as those shown above in the design of organic syntheses, we need to remember which kinds of reactants give which kinds of products. For example, we need to remember that when bromine reacts with propene, the product is 1,2-dibromopropane. Rather than learning each reaction as an isolated fact, however, it is important to recognize general patterns that allow us to predict with some certainty how a particular set of compounds will react, even though we may not have encountered the specific compounds before[4]. For example, if we know how propene reacts with bromine, we can correctly predict that cyclohexene will react with chlorine to form 1,2-dichlorocyclohexane. In fact, we can rely on this type of reaction as a general diagnostic test for the presence of a carbon-carbon multiple bond in a molecule. Thus, if an unknown compound is treated with a dilute solution of bromine in carbon tetrachloride (pale brown-red color) and the solution becomes colorless, it is a reasonable conclusion that the compound contains a carbon-carbon multiple bond to which the bromine in the solution has added. However, in order to make predictions and extrapolations of this kind, it is useful to have some knowledge of how the reactions take place, i.e., the mechanisms by which they occur. Although it can be risky to have absolute trust in all the generalizations of organic chemistry, some can still be helpful to those encountering the subject for the first time. Despite the exceptions, generalizations provide very useful focal points around which to organize one's thoughts. Recognizing the potential pitfalls, we suggest the following:

(1) Saturated alkanes exhibit radical substitution as their primary reaction, while alkenes and alkynes display both ionic and radical addition reactions. Conversely, aromatic compounds undergo electrophilic substitution in their characteristic reactions.

(2) The relative reactivity of C—H bonds in terms of hydrogen atom abstraction decreases in the following order:

$$C=C-\overset{|}{\underset{|}{C}}-H \approx Ar-\overset{|}{\underset{|}{C}}-H > R_3CH > R_2CH_2 > RCH_3 > CH_4$$

(3) The resistance of different carbon backbones decreases in the following order: arene > alkane > alkyne > alkene. However, this must be accepted with reservations due to the reaction conditions and the fact that some oxidants, e.g., potassium permanganate, oxidizes the methyl group rather than the benzene ring, whereas ozone preferentially attacks the benzene ring.

The compounds whose reactions are discussed above are those composed of carbon and hydrogen alone. However, we have already seen that heteroatoms and groups can be attached to carbon skeletons, and it is not surprising that the chemistry of the original carbon skeleton can be changed as a result. For example, chloroform ($CHCl_3$) has a hydrogen that is sufficiently acidic to be removed by strong bases; tetrafluoroethylene ($CF_2=CF_2$) is inert to electrophilic addition and instead reacts readily with nucleophiles; benzoic acid (C_6H_5COOH) is much less susceptible to electrophilic substitution than benzene; and *p*-hydroxytoluene ($HO-C_6H_4-CH_3$) undergoes oxidation in the aromatic ring rather than at the methyl group. Thus, the attached groups (i.e., the functional groups) can alter the properties of the carbon backbones to the point of reversing the chemical behavior observed in the parent systems.

5.3 Name Reactions in Organic Chemistry

There are several important named reactions in organic chemistry, so called either because they bear the names of the people who described them, or because they are referred to by a particular name in texts and journals. Sometimes the name gives a clue to the reactants and products, but not always. Here are some of the names and equations for the most important reactions, listed in alphabetical order.

5.3.1 Acetoacetic-Ester Condensation Reaction

The acetoacetate condensation reaction converts a pair of ethyl acetate ($CH_3COOC_2H_5$) molecules into ethyl acetoacetate ($CH_3COCH_2COOC_2H_5$) and ethanol (CH_3CH_2OH) in the presence of sodium ethoxide (NaOEt) and hydronium ions (H_3O^+).

5.3.2 Acetoacetic Ester Synthesis

In this organic name reaction, the acetoacetic ester synthesis reaction converts an α-keto acetic acid to a ketone. The most acidic methylene group reacts with the base and adds the alkyl group in its place. The product of this reaction can be treated again with the same or different alkylating agent (the downward reaction) to create a dialkyl product.

5.3.3 Acyloin Condensation

In the acyloin condensation reaction, two carboxylic acid esters are condensed in the presence of sodium metal to form an α-hydroxyketone, also known as acyloin. Intramolecular acyloin condensation can be used for ring closure, as in the second reaction.

5.3.4 Alder-Ene Reaction or Ene Reaction

The Alder-ene reaction, also known as the ene reaction, is a group reaction that combines an ene and an enophile. The ene is an alkene with an allylic hydrogen and the enophile is a multiple bond. The reaction produces an alkene in which the double bond is moved to the allylic position.

5.3.5 Aldol Reaction or Aldol Addition

The aldol addition reaction is the combination of an alkene or ketone with the carbonyl of another aldehyde or ketone to form a β-hydroxyaldehyde or ketone. Aldol is a combination of the terms "aldehyde" and "alcohol".

5.3.6 Aldol Condensation Reaction

Aldol condensation removes the hydroxyl group formed by the aldol addition reaction in the form of water in the presence of an acid or base. Aldol condensation forms α,β-unsaturated carbonyl compounds.

5.3.7 Appel Reaction

The Appel reaction converts an alcohol to an alkyl halide using triphenylphosphine (PPh_3) and either tetrachloromethane (CCl_4) or tetrabromomethane (CBr_4).

5.3.8 Arbuzov Reaction or Michaelis-Arbuzov Reaction

The Arbuzov or Michaelis-Arbuzov reaction combines a trialkyl phosphate with an alkyl halide (the X in the reaction is a halogen) to form an alkyl phosphonate.

5.3.9 Arndt-Eistert Synthesis Reaction

The Arndt-Eistert synthesis is a series of reactions to make a carboxylic acid homologue. In this synthesis, a carbon atom is added to an existing carboxylic acid.

5.3.10 Azo Coupling Reaction

The azo coupling reaction combines diazonium ions with aromatic compounds to form azo compounds. Azo coupling is commonly used to make pigments and dyes.

5.3.11 Baeyer-Villiger Oxidation Reactions

The Baeyer-Villiger oxidation reaction converts a ketone to an ester. This reaction requires the presence of a peracid such as mCPBA or peroxyacetic acid. Hydrogen peroxide can be used in conjunction with a Lewis base to form a lactone ester.

5.3.12 Baker-Venkataraman Rearrangement

The Baker-Venkataraman rearrangement reaction converts an ortho-acylated phenol ester to a 1,3-diketone.

5.3.13 Balz-Schiemann Reaction

The Balz-Schiemann reaction is a process for converting aryl amines to aryl fluorides by diazotization.

5.3.14 Bamford-Stevens Reaction

The Bamford-Stevens reaction converts tosylhydrazones to alkenes in the presence of a strong base. The nature of the alkene depends on the solvent used. Protic solvents produce carbenium ions and aprotic solvents produce carbene ions.

5.3.15 Barton Decarboxylation

In the Barton decarboxylation reaction, a carboxylic acid is converted to a thiohydroxamate ester, commonly called a Barton ester, and then reduced to the corresponding alkane.

DCC = N,N'-dicyclohexylcarbodiimide
DMAP = 4-dimethylaminopyridine
AIBN = 2,2'-azobisisobutyronitrile

5.3.16 Barton Deoxygenation Reaction: Barton-McCombie Reaction

The Barton deoxygenation reaction removes the oxygen from alkyl alcohols. The hydroxy group is replaced with a hydride to form a thiocarbonyl derivative, which is then treated with Bu_3SnH, which removes all but the desired radical.

R = alkyl
R' = H, CH_3, SCH_3, OCH_3, Ph, imidazolyl

5.3.17 Baylis-Hillman Reaction

The Baylis-Hillman reaction combines an aldehyde with an activated alkene. This reaction is catalyzed by a tertiary amine molecule such as DABCO (1, 4-diazabicyclo[2. 2. 2]octane). EWG is an electron withdrawing group that removes electrons from aromatic rings.

5.3.18 Beckmann Rearrangement Reaction

The Beckmann rearrangement reaction converts oximes to amides. Cyclic oximes give rise to lactam molecules.

5.3.19 Benzilic Acid Rearrangement

The benzilic acid rearrangement reaction rearranges a 1, 2-diketone to an α-hydroxycarboxylic acid in the presence of a strong base. Cyclic diketones contract the ring by benzilic acid rearrangement.

5.3.20 Benzoin Condensation Reaction

The benzoin condensation reaction condenses a pair of aromatic aldehydes to an α-hydroxyketone.

5.3.21 Bergman Cycloaromatization—Bergman Cyclization

In Bergman cycloaromatization, also known as Bergman cyclization, enediynes are formed from substituted arenes in the presence of a proton donor such as 1,4-cyclohexadiene. This reaction can be initiated by either light or heat.

5.3.22 Bestmann-Ohira Reagent Reaction

The Bestmann-Ohira reaction is a special case of the Seyferth-Gilbert homologation reaction. The Bestmann-Ohira reagent uses dimethyl-1-diazo-2-oxopropylphosphonate to form alkynes from an aldehyde.

THF = tetrahydrofuran

5.3.23 Biginelli Reaction

The Biginelli reaction combines ethyl acetoacetate, an arylaldehyde and urea to form dihydropyrimidones (DHPMs). The arylaldehyde in this example is benzaldehyde.

5.3.24 Birch Reduction Reaction

The Birch reduction reaction converts aromatic compounds with benzenoid rings to 1,4-cyclohexadienes. The reaction takes place in ammonia, alcohol and in the presence of sodium, lithium or potassium.

5.3.25 Bicschler-Napieralski Reaction — Bicschler-Napieralski Cyclization

The Bicschler-Napieralski reaction produces dihydroisoquinolines by the cyclization of β-ethylamides or β-ethylcarbamates.

5.3.26 Blaise Reaction

The Blaise reaction combines nitriles and α-haloesters using zinc as a mediator to form β-enamino esters or β-keto esters. The form produced depends on the addition of the acid.

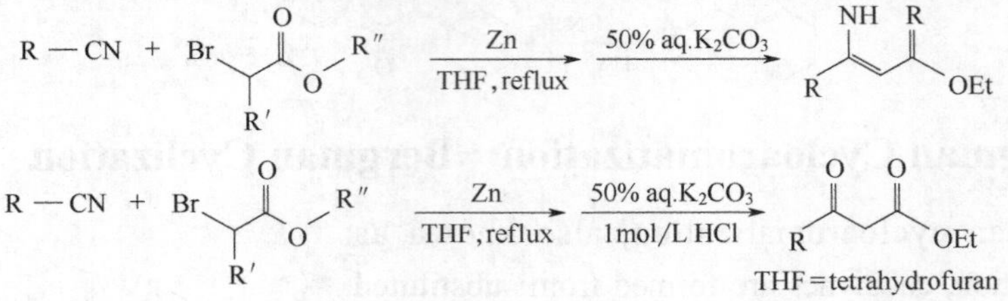

$$R-CN + Br-CR'(R'')C(O)O-R'' \xrightarrow[THF, reflux]{Zn} \xrightarrow{50\% \ aq.K_2CO_3} \text{β-enamino ester}$$

$$R-CN + Br-CR'(R'')C(O)O-R'' \xrightarrow[THF, reflux]{Zn} \xrightarrow[1mol/L \ HCl]{50\% \ aq.K_2CO_3} \text{β-keto ester}$$

THF = tetrahydrofuran

5.3.27 Blanc Reaction

The Blanc reaction produces chloromethylated arenes from an arene, formaldehyde, HCl, and zinc chloride. If the concentration of the solution is high enough, a secondary reaction occurs with the product and the arenes.

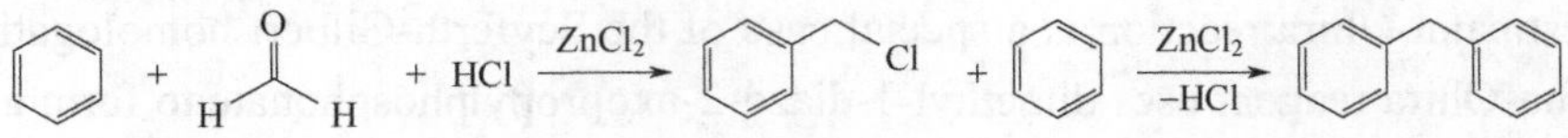

5.3.28 Bohlmann-Rahtz Pyridine Synthesis

The Bohlmann-Rahtz pyridine synthesis produces substituted pyridines by condensing enamines and ethynyl ketones to form an aminodiene and then a 2, 3, 6-trisubstituted pyridine. The EWG group is an electron withdrawing group.

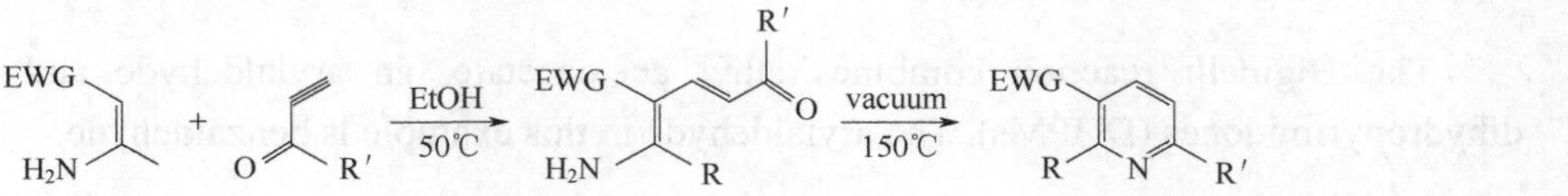

5.3.29 Bouveault-Blanc Reduction

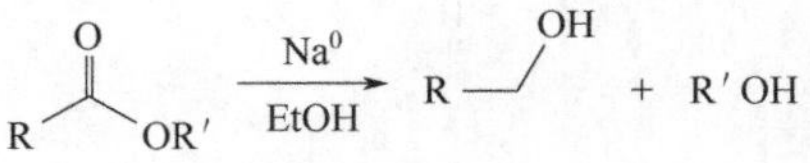

The Bouveault-Blanc reduction reduces esters to alcohols in the presence of ethanol and sodium metal.

5.3.30 Brook Rearrangement

The Brook rearrangement transports the silyl group on an α-silyl carbinol from a carbon to the oxygen in the presence of a base catalyst.

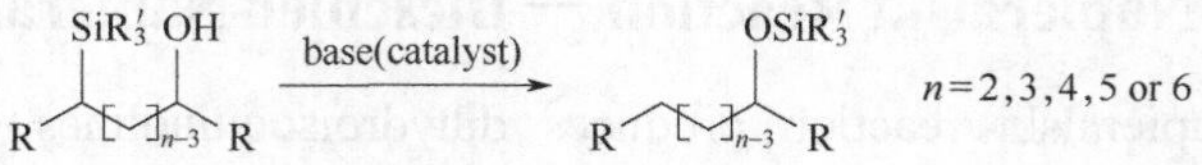

$n = 2, 3, 4, 5$ or 6

5.3.31 Brown Hydroboration

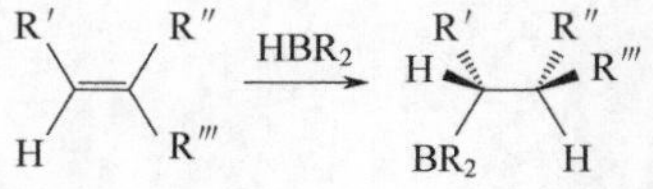

The Brown hydroboration reaction combines hydroborane compounds to alkenes. The boron binds to the least hindered carbon.

5.3.32 Bucherer-Bergs Reaction

In the Bucherer-Bergs reaction, a ketone, potassium cyanide, and ammonium carbonate combine to form hydantoins. In the second reaction, a cyanohydrin and ammonium carbonate form the same product.

5.3.33 Buchwald-Hartwig Cross Coupling Reaction

The Buchwald-Hartwig cross-coupling reaction forms arylamines from aryl halides or pseudohalides and primary or secondary amines using a palladium catalyst. The second reaction shows the synthesis of aryl ethers using a similar mechanism.

R=alkyl,CN,COR;X=halogen;R′=alkyl,aryl

Ar,Ar′=aryl;X=halogen

5.3.34 Cadiot-Chodkiewicz Coupling Reaction

The Cadiot-Chodkiewicz coupling reaction produces bisacetylenes from the combination of a terminal alkyne and an alkynyl halide using a copper(I) salt as a catalyst.

5.3.35 Cannizzaro Reaction

The Cannizzaro reaction is a redox disproportionation of aldehydes to carboxylic acids and alcohols in the presence of a strong base. The second reaction uses a similar mechanism with α-keto aldehydes. The Cannizzaro reaction sometimes produces unwanted by-products in reactions involving aldehydes under basic conditions.

5.3.36 Chan-Lam Coupling Reaction

The Chan-Lam coupling reaction forms aryl carbon-heteroatom bonds by combining arylboronic compounds, stannanes or siloxanes with compounds containing either an N—H or O—H bond. The reaction uses a copper catalyst that can be reoxidized by oxygen in the air at room temperature. Substrates can include amines, amides, anilines, carbamates, imides, sulfonamides and ureas.

$$\text{Ar—B(OH)}_2 + \text{YH—R} \xrightarrow[\text{CH}_2\text{Cl}_2,\ \text{RT}]{\text{Cu(OAc)}_2} \text{Ar—Y—R} \quad \text{Y=NR}',\ \text{O, S, NCOR}'\ \text{or NSO}_2\text{R}'$$

5.3.37 Crossed Cannizzaro Reaction

The crossed Cannizzaro reaction is a variant of the Cannizzaro reaction where formaldehyde is the reducing agent.

$$\text{PhCHO} + \text{H}_2\text{C=O} \xrightarrow{\text{NaOH}} \text{PhCH}_2\text{OH} + \text{HCOOH}$$

5.3.38 Friedel-Crafts Reaction

A Friedel-Crafts reaction is the alkylation of benzene. When a haloalkane is reacted with benzene using a Lewis acid (usually an aluminum halide) as a catalyst, the alkane is attached to the benzene ring and excess hydrogen halide is produced. This is also known as Friedel-Crafts alkylation of benzene.

$$\text{RX} + \text{C}_6\text{H}_6 \xrightarrow{\text{AlCl}_3} \text{C}_6\text{H}_5\text{R} + \text{HX}$$

5.3.39 Huisgen Azide-Alkyne Cycloaddition Reaction

The Huisgen azide-alkyne cycloaddition combines an azide compound with an alkyne compound to form a triazole compound. The first reaction uses only heat to form 1,2,3-triazoles. The second reaction uses a copper catalyst and forms only 1,4-disubstituted 1,2,3-triazoles. The third reaction uses a ruthenium and cyclopentadienyl (Cp) compound as a catalyst to form 1,5-disubstituted 1,2,3-triazoles.

$$\text{R—N}_3 + \text{≡—R}' \begin{cases} \xrightarrow{\text{heat}} & \text{1,4- and 1,5-triazoles} \\ \xrightarrow[\text{H}_2\text{O}]{\text{Cu(I) catalyst}} & \text{1,4-disubstituted triazole} \\ \xrightarrow[\text{heat}]{\text{Cp*RuCl(PPh}_3\text{) catalyst}} & \text{1,5-disubstituted triazole} \end{cases}$$

5.3.40 Itsuno-Corey Reduction — Corey-Bakshi-Shibata Reduction

The Itsuno-Corey reduction, also known as the Corey-Bakshi-Shibata (CBS) reduction, is an enantioselective reduction of ketones in the presence of a chiral oxazaborolidine (CBS) catalyst and borane.

5.3.41 Seyferth-Gilbert Homologation Reaction

The Seyferth-Gilbert homologation reacts aldehydes and aryl ketones with dimethyl (diazomethyl)phosphonate to synthesize alkynes at low temperatures.

New Words and Expressions

originally	[əˈrɪdʒənəli]	adv. 起初，原来，独创地
urea	[juˈriːə]	n. 尿素
starch	[stɑːtʃ]	n. 淀粉，含淀粉的食物，v. 给……上浆
wax	[wæks]	n. 蜡，蜂蜡，石蜡
vitalism	[ˈvaɪtəˌlɪzəm]	n. 生命力
mineral	[ˈmɪnərəl]	n. 矿物质，矿物，adj.（与）矿物（有关）的
molecule	[ˈmɒlɪkjuːl]	n. 分子，摩尔，少量，微粒
vet	[vet]	n. 兽医；v. 仔细检查，诊疗
pharmacologist	[ˌfɑːməˈkɒlədʒɪst]	n. 药理学家
plastic	[ˈplæstɪk]	n. 塑料，塑料制品
synthesize	[ˈsɪnθəsaɪz]	v.（通过化学或生物反应）合成
ammonia	[əˈməʊniə]	n. 氨
absence	[ˈæbsəns]	n. 缺席，缺勤，没有
lack	[læk]	n. 缺乏，不足；v. 缺乏，不足，需要
contaminate	[kənˈtæmɪneɪt]	v. 污染，弄脏，毒害，腐蚀
herbal	[ˈhɜːbl]	adj. 药草的，香草的；n. 草本植物志
flavoring	[ˈfleɪvərɪŋ]	n. 调味品
artificial	[ˌɑːtɪˈfɪʃ(ə)l]	adj. 人造的，人工的，人为的
decay	[dɪˈkeɪ]	v.（建筑）破败，衰落，（观念）衰败；n. 腐烂，衰退
radioactive	[ˌreɪdiəʊˈæktɪv]	adj. 放射性的，有辐射的
isotope	[ˈaɪsətəʊp]	n. 同位素

sophisticated	[səˈfɪstɪkeɪtɪd]	adj.	见多识广的，先进的，精密的，水平高的
substance	[ˈsʌbstəns]	n.	物质，材料，主旨
graphite	[ˈɡræfaɪt]	n.	石墨
dioxide	[daɪˈɒksaɪd]	n.	二氧化物
element	[ˈelɪmənt]	n.	基本部分，要素，元素
abundant	[əˈbʌndənt]	adj.	大量的，丰富的，充足的
crust	[krʌst]	n.	地壳
nourish	[ˈnʌrɪʃ]	v.	养育，滋养，培养
protein	[ˈprəʊtiːn]	n.	蛋白质，朊
glycogen	[ˈɡlaɪkədʒən]	n.	糖原，动物淀粉
pesticide	[ˈpestɪsaɪd]	n.	杀虫剂，除害药物
fiber	[ˈfaɪbə(r)]	n.	纤维
organ	[ˈɔːɡən]	n.	器官
heteroatom	[ˈhetərəʊˌætəm]	n.	杂原子，杂环原子
correct	[kəˈrekt]	adj.	正确的，无误的；v. 改正，纠正，批改
appreciation	[əˌpriːʃɪˈeɪʃn]	n.	欣赏，感激，理解，增值
framework	[ˈfreɪmwɜːk]	n.	构架，结构，参照标准
ornament	[ˈɔːnəmənt]	n.	装饰品，点缀品；v. 装饰，点缀
ethane	[ˈiːθeɪn]	n.	乙烷
dissolve	[dɪˈzɒlv]	v.	（以化学手段）除去，分散，（使）溶解
ingredient	[ɪnˈɡriːdiənt]	n.	成分，原料，要素，因素；adj. 构成组成部分的
beverage	[ˈbevərɪdʒ]	n.	饮料
nitrogen	[ˈnaɪtrədʒən]	n.	氮
pungent	[ˈpʌndʒənt]	adj.	辛辣的，刺激性的
fuel	[ˈfjuːəl]	n.	燃料，燃烧剂
mechanism	[ˈmekənɪzəm]	n.	机械装置，途径，方法
generate	[ˈdʒenəreɪt]	v.	产生，引起
elimination	[ɪˌlɪmɪˈneɪʃn]	n.	消除，排除
carbonium	[kɑːˈbəʊniəm]	n.	碳正离子
bromide	[ˈbrəʊmaɪd]	n.	溴化物
reagent	[riˈeɪdʒənt]	n.	试剂，反应物
propene	[ˈprəʊpiːn]	n.	丙烯
isopropyl	[ˌaɪsəʊˈprəʊpɪl]	n.	异丙基
proton	[ˈprəʊtɒn]	n.	质子
chlorine	[ˈklɔːriːn]	n.	氯
irradiate	[ɪˈreɪdieɪt]	v.	启发，放射，照射；adj. 发光的
dissociation	[dɪˌsəʊʃiˈeɪʃn]	n.	分解
collide	[kəˈlaɪd]	v.	冲突，碰撞，相撞
butyl	[ˈbjuːtɪl]	n.	丁基

terminate	[ˈtɜ:mɪneɪt]	v.（使）结束，（使）终止
saturated	[ˈsætʃəreɪtɪd]	adj.（溶液）饱和的
alicyclic	[ˌælɪˈsaɪklɪk]	n. 脂环族
platinum	[ˈplætɪnəm]	n. 铂，白金
palladium	[pəˈleɪdiəm]	n. 钯
selenium	[səˈli:niəm]	n. 硒
tetralin	[ˈtetrəlɪn]	n. 萘满，1, 2, 3, 4-四氢化萘
decalin	[ˈdekəlɪn]	n. 十氢化萘，萘烷
naphthalene	[ˈnæfθəli:n]	n. 萘，卫生球
aromatic	[ˌærəˈmætɪk]	adj. 芳香的，芳香族的；n. 芳香植物，芳香剂
crystallizable	[krɪstəˈlaɪzəbl]	adj. 可结晶的
stereochemically	[steri:əuˈkemɪklɪ]	adv. 用立体化学方法
precursor	[prɪˈkɜ:sə(r)]	n. 先兆，前体，前兆
cleavage	[ˈkli:vɪdʒ]	n. 劈开，解理
ozone	[ˈəuzəun]	n. 臭氧
carboxylic	[ˌkɑ:bɒkˈsɪlɪk]	adj. 羧基的
1, 2-dibromopropane	[daɪbrəuməpˈrɒpeɪn]	n. 1, 2-二溴丙烷
cyclohexene	[ˈsaɪkləuhɪksi:n]	n. 环己烯
diagnostic	[ˌdaɪəgˈnɒstɪk]	adj. 诊断的，判断的
tetrachloride	[ˌtetrəˈklɔ:raɪd]	n. 四氯化物
extrapolation	[ɪkˌstræpəˈleɪʃn]	n. 外推法，推断
abstraction	[æbˈstrækʃn]	n. 抽象，提取
resistance	[rɪˈzɪstəns]	n. 反对，阻力，电阻
potassium	[pəˈtæsiəm]	n. 钾
permanganate	[pəˈmæŋgənət]	n. 高锰酸盐
benzene	[ˈbenzi:n]	n. 苯
chloroform	[ˈklɒrəfɔ:m]	n. 氯仿，三氯甲烷
tetrafluoroethylene	[ˌtetrəˌfluərəuˈeθɪli:n]	n. 四氟乙烯
electrophilic	[ɪˌlektrəuˈfɪlɪk]	adj. 亲电子的
benzoic acid		苯甲酸
p-hydroxytoluene		n. 对羟基甲苯
acetoacetate	[æsɪtəˈæsɪteɪt]	n. 乙酰乙酸盐
ethyl acetate		乙酸乙酯
ethyl acetoacetate		乙酰乙酸乙酯
ethanol	[ˈeθənɒl]	n. 乙醇，酒精
ethoxide	[i:ˈθɒksaɪd]	n. 乙醇盐
alkylation	[ˌælkɪˈleɪʃən]	n. 烃化，烷化
acyloin	[ˈeɪsaɪˌəuɪn]	n. 酮醇，偶姻
hydroxyketone	[haɪˌdrɒksɪˈki:təun]	n. 羟基酮，醇酚
enophile	[ˈi:nəfaɪl]	n. 亲烯物
ketone	[ˈki:təun]	n. 酮

carbonyl	[ˈkɑːbəˌnaɪl]	n.	羰基
hydroxyaldehyde	[haɪdrɒksiˈæːldɪhaɪd]	n.	羟醛
aldehyde	[ˈældɪhaɪd]	n.	醛
unsaturate	[ʌnˈsætʃəreɪt]	n.	不饱和化合物
halogen	[ˈhælədʒən]	n.	卤素
phosphonate	[ˈfɒsfəuneɪt]	n.	膦酸酯，膦酸盐
pigment	[ˈpɪgmənt]	n.	色素，颜料
dye	[daɪ]	n.	染料，染液
ester	[ˈestə(r)]	n.	酯
aryl	[ˈærɪl]	n.	芳香基
fluoride	[ˈflɔːraɪd]	n.	氟化物
tosylhydrazone	[təuzɪlaɪdreɪˈzəun]	n.	甲苯磺酰腙
deoxygenation	[diːɒksɪdʒəˈneɪʃən]	n.	脱氧，脱氧作用
thiocarbonyl	[θaɪouˈkɑːbənɪl]	n.	硫代羰基
catalyze	[ˈkætəlaɪz]	v.	催化，刺激
condense	[kənˈdens]	v.	冷凝，凝结，压缩
cyclization	[ˌsaɪkləˈzeɪʃən]	n.	环化，环合作用
dimethyl	[daɪˈmiːθaɪl]	n.	乙烷，二甲基
oxopropylphosphonate		n.	膦酸氧丙酯
chloromethylate		n.	氯甲基化
pyridine	[ˈpɪrɪdiːn]	n.	吡啶
aminodiene		n.	氨基二烯
catalyst	[ˈkætəlɪst]	n.	催化剂
carbonate	[ˈkɑːbənət]	n.	碳酸盐
bisacetylene		n.	双乙炔
copper	[ˈkɒpə(r)]	n.	铜
triazole	[ˈtraɪəˌzɒl]	n.	三唑
borane	[ˈbɔːreɪn]	n.	硼烷

Notes

1. Although carbon is not one of the ten most abundant elements in the Earth's crust, surface waters and atmosphere, it is the third most abundant element in the human body in terms of number of atoms and the second most abundant in terms of mass.
虽然碳不是地壳、地表水和大气中含量最高的十大元素之一，但就原子数量而言，它是人体中含量第三高的元素，就质量而言，它是含量第二高的元素。

2. The importance of the combustion reaction cannot be minimized: most of the heat and energy (apart from that generated by falling water, wind, sun, compressed steam, and within the Earth or by nuclear sources) comes from the combustion of very complicated oxidative processes in which carbon compounds are ultimately converted to carbon dioxide and water.
燃烧反应的重要性不言而喻：大部分热量和能量（除了由流水、风、太阳、压缩蒸汽和地球内部或核源产生的热量和能量）都来自非常复杂的氧化燃烧过程，在这个过程中，碳化合物最终转化为二氧化碳和水。

3. Once this stage of the reaction sequence is reached, the process becomes self-perpetuating; each time a carbon radical extracts a hydrogen from hydrogen bromide, another bromine atom is formed, which adds itself to propene to form another carbon radical.

一旦达到反应序列的这一阶段，反应过程就会自我循环；每当一个碳自由基从溴化氢中获取一个氢时，就会形成另一个溴原子，而这个溴原子又会与丙烯结合，形成另一个碳自由基。

4. Rather than learning each reaction as an isolated fact, however, it is important to recognize general patterns that allow us to predict with some certainty how a particular set of compounds will react, even though we may not have encountered the specific compounds before.

然而，重要的不是把每个反应作为一个孤立的事实来学习，而是要认识到一般的模式，使我们能够比较肯定地预测一组特定的化合物将如何反应，即使我们以前可能没有遇到过这些化合物。

Trivial — Last but not the Least

Geometry Formulas

1. 正方形(square)

(1) 已知边长求面积　　area of square = side squared
$S = a^2$　　　　　　　the area of a square equals the side squared

(2) 求周长　　　　　　perimeter of square = side × 4
$C = 4a$　　　　　　　the perimeter of a square equals the side times four

(3) 已知面积求边长　　side = square root of the area
$a = \sqrt{S}$　　　　　the side of a square equals the square root of the area

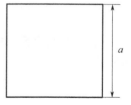

2. 长方形(rectangle)

(1) 已知边长求　　　　area of rectangle = length × height
面积 $S = h \times l$　　the area of a rectangle equals the length of the rectangle times the height

(2) 已知面积和　　　　length of rectangle = area ÷ height
高求边长 $l = S/h$　　the length of rectangle equals the area divided by the height

(3) 已知面积和　　　　height of rectangle = area ÷ length
边长求高 $h = S/l$　　the height of a rectangle equals the area divided by the length

(4) 已知边长和　　　　perimeter of rectangle = 2(length + height)
高求周长　　　　　　the perimeter of a rectangle equals two open
$C = 2(l + h)$　　　　brackets the length plus the height close brackets

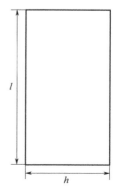

3. 三角形(triangle)

(1) 已知底和高求面积
$S =(a h)/2$

area of triangle = ½ (base × height)
the area of a triangle equals the base of the triangle multiplied by the height divided by two

(2) 已知面积和底边长度求高
$h = 2S/a$

height of triangle = (area × 2) ÷ base
the height of a triangle equals the area times two divided by the base

(3) 已知面积和高求底 $a = 2S/h$

base of triangle = (area × 2) ÷ height
the base of a triangle equals the area times two divided by the height

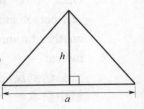

4. 圆形(circle)

(1) 已知半径求周长 $C = 2\pi R$

circumference of a circle = twice radius × π
the circumference of a circle equals twice radius times π (pi)

(2) 已知周长求半径 $R = C/(2\pi)$

radius = circumference of circle ÷ 2π
the radius of a circle equals the circumference divided by two π (pi)

(3) 已知半径求面积 $S =\pi R^2$

area = radius squared multiplied by π (pi)
the area of a circle equals the radius squared multiplied by π (pi)

(4) 已知周长求面积 $S = C^2/(4\pi)$

area = circumference squared over four π (pi)
the area of a circle equals the square of the circumference divided by four times π (pi)

掌握以上简单公式中所包含的英文表达，可以为今后进一步学习比较复杂的公式打下初步的基础。

Unit 6

Rules of Nomenclature of Organic Chemistry

Although this unit covers a wide range of related organic compounds, it only briefly introduces and names a large number of different types of organic compounds. The nomenclature of organic compounds presented in this unit will most likely be used repeatedly in future work related to materials chemistry.

6.1 Hydrocarbons

The term "hydrocarbon" refers to compounds consisting only of carbon and hydrogen atoms. Aliphatic hydrocarbons consist of chains of carbon atoms that do not have a cyclic structure. They are often referred to as open chain or acyclic hydrocarbons. Alicyclic, or simply cyclic, hydrocarbons are composed of carbon atoms arranged in one or more rings. Aromatic hydrocarbons are a special group of cyclic compounds that typically have six-membered rings with alternating single and double bonds. They are classified separately from alicyclic and aliphatic hydrocarbons because of their characteristic physical and chemical properties.

6.1.1 Alkanes

Alkanes are aliphatic or alicyclic hydrocarbons in which each carbon atom is bonded to four other carbon atoms. Alkanes are also known as paraffins or saturated hydrocarbons. We have already learned that saturated aliphatic hydrocarbons have the empirical formula C_nH_{2n+2}. The term cycloalkane is often used to describe saturated cyclic hydrocarbons. Monocyclic alkanes have the empirical formula C_nH_{2n}.

When the carbon atoms of an alkane are linked in a single continuous sequence, the alkane is known as a normal hydrocarbon. These alkanes are sometimes called linear or straight chain hydrocarbons, but these names are misleading. In reality the carbon chains are bent and twisted and are only linear in our symbolic representations.

A homologous series refers to a set of hydrocarbons that differ in the number of methylene groups (—CH_2—). Example includes methane (CH_4), ethane (C_2H_6), propane (C_3H_8), and butane (C_4H_{10}). The boiling points of these homologues tend to increase in a regular way as the carbon chain length increases.

The basic criterion for naming a structure in the IUPAC (International Union of Pure and

Applied Chemistry) system is the choice of a parent name. The longest continuous sequence of carbon atoms in the molecule is the basis for the parent name of an aliphatic hydrocarbon. The parent name is derived from a root word, typically of Greek origin, which indicates the number of atoms in the longest continuous chain. For example, a sequence of five atoms is named with the root word pent-, while a sequence of the ten atoms is named with the root word dec-. The -ane suffix is added to the stem to complete the parent name for an alkane. The name of a normal alkane is identical to its parent name. The roots and corresponding names of normal alkanes are listed in Table 6.1 below.

Table 6.1 Roots and related names of normal alkanes

Number of carbon atoms in longest continuous chain	Root	Condensed formula of normal alkane	Name
1	meth-	CH_4	methane
2	eth-	CH_3CH_3	ethane
3	prop-	$CH_3CH_2CH_3$	propane
4	but-	$CH_3(CH_2)_2CH_3$	butane
5	pent-	$CH_3(CH_2)_3CH_3$	pentane
6	hex-	$CH_3(CH_2)_4CH_3$	hexane
7	hept-	$CH_3(CH_2)_5CH_3$	heptane
8	oct-	$CH_3(CH_2)_6CH_3$	octane
9	non-	$CH_3(CH_2)_7CH_3$	nonane
10	dec-	$CH_3(CH_2)_8CH_3$	decane
11	undec-	$CH_3(CH_2)_9CH_3$	undecane
12	dodec-	$CH_3(CH_2)_{10}CH_3$	dodecane
13	tridec-	$CH_3(CH_2)_{11}CH_3$	tridecane
14	tetradec-	$CH_3(CH_2)_{12}CH_3$	tetradecane
15	pentadec-	$CH_3(CH_2)_{13}CH_3$	pentadecane
16	hexadec-	$CH_3(CH_2)_{14}CH_3$	hexadecane
17	heptadec-	$CH_3(CH_2)_{15}CH_3$	heptadecane
18	octadec-	$CH_3(CH_2)_{16}CH_3$	octadecane
19	nonadec-	$CH_3(CH_2)_{17}CH_3$	nonadecane
20	eicosan-	$CH_3(CH_2)_{18}CH_3$	eicosane

Note that halogen substituents may also be designated by prefixes like fluoro (F—), chloro (Cl—), bromo (Br—), and iodo (I—). The substituent's location is signified by numbers. For example, $(CH_3)_2CHCH_2CH_2Br$ represents 1-bromo-3-methylbutane.

For compounds more complex than normal hydrocarbons, the name of the normal hydrocarbon derived from the longest chain of carbon atoms in the compound is the parent name upon which further nomenclature is based[1]. A secondary prefix is used to denote the structure of side chain groups attached to the parent chain. Hydrocarbon side chain structures are named using the appropriate root name for that fragment with a -yl as the suffix. For example, a CH_3— group attached to the parent chain is called methyl, a CH_3CH_2— group is called ethyl, and so on. Alkyl is the general term for an attached hydrocarbon group. The position of attachment of a side chain group is indicated by numbering the parent chain from one end so that the atom to which the side chain is attached is given the lowest possible number. The number immediately precedes the name of the side chain to which it applies. Note the convention that numbers are separated from words by a hyphen and that words are written without spaces between their parts.

$$\underset{1\ \ 2\ \ 3\ \ 4\ \ 5\ \ 6}{CH_3\overset{\overset{CH_3}{|}}{C}HCH_2CH_2CH_2CH_3}$$

2-methylhexane

$$\underset{1\ \ 2\ \ 3\ \ 4\ \ 5\ \ 6\ \ 7}{CH_3CH_2CH_2\overset{\overset{CH_2CH_3}{|}}{C}HCH_2CH_2CH_3}$$

4-ethylheptane

Multiple side chain groups are labeled with the appropriate Greek multiplicative numerical prefix, i.e., di- for two, tri- for three, etc., and numbered starting with the lowest possible number. Here we see in practice the conventions that each substituent group on the longest chain must be designated by a number, that the numbers are separated by commas, and that the lowest possible numbers are used.

$$\underset{1\ \ 2\ \ 3\ \ 4\ \ 5}{CH_3CH_2\overset{\overset{CH_3}{|}}{\underset{\underset{CH_3}{|}}{C}}CH_2CH_3}$$

3, 3-dimethylpentane

$$\underset{1\ \ 2\ \ 3\ \ 4\ \ 5\ \ 6\ \ 7}{CH_3\overset{\overset{CH_3}{|}}{C}HCH_2\overset{\overset{CH_3}{|}}{C}HCH_2CH_2CH_3}$$

2, 4-dimethylheptane (NOT 4, 6-dimethylheptane)

$$\underset{1\ \ 2\ \ 3\ \ 4\ \ 5\ \ 6}{CH_3\overset{\overset{CH_3}{|}}{C}H-\overset{\overset{CH_3}{|}}{C}HCH_2\overset{\overset{CH_3}{|}}{C}HCH_3}$$

2, 3, 5-trimethylhexane (NOT 2, 4, 5-trimethylhexane)

When two or more different substituent groups are attached to the parent chain, the names of these groups are arranged alphabetically in the compound name, regardless of their number. The IUPAC rules are flexible on this point, since the alphabetical order of the group names may be different in different languages. If two different numbering sequences begin with the same number, the sequence with the lower number is selected at the first point of difference. Please note that this is different from the order in the "Nomenclature of Organic Compounds" established by the Chinese Chemical Society.

$$\underset{1\ \ 2\ \ 3\ \ 4\ \ 5\ \ 6\ \ 7\ \ 8\ \ 9}{CH_3CH_2\overset{\overset{CH_3}{|}}{C}HCH_2\overset{\overset{CH_2CH_2CH_3}{|}}{C}HCH_2CH_2CH_2CH_3}$$

3-methyl-5-propylnonane

$$\underset{1\ \ 2\ \ 3\ \ 4\ \ 5\ \ 6}{CH_3\overset{\overset{CH_3}{|}}{C}HCH_2\overset{\overset{CH_2CH_3}{|}}{C}HCH_2CH_3}$$

4-ethyl-2-methylhexane

In addition to the systematically named straight-chain examples above, alkanes have branched structures. The smallest hydrocarbon that can form a branched chain has four carbon atoms. Such compounds have the same chemical formula as butane (C_4H_{10}), but different structures. Compounds with the same formula and different structures are known as isomers (from the Greek isos, "equal", and meros, "parts"). When it was first discovered, the branched isomer with the formula C_4H_{10} was therefore given the name isobutane. Butane and isobutane are called constitutional isomers because they literally differ in their constitution. One contains two CH_3 groups and two CH_2 groups; the other contains three CH_3 groups and one CH group. There are three constitutional isomers of pentane, C_5H_{12}. The first is normal pentane, or *n*-pentane. A branched isomer is also possible, which was originally named isopentane. When a more highly branched isomer was discovered, it was named neopentane (the new isomer of pentane). The conventional nomenclature for naming alkanes with the prefix "n" "iso", and "neo" is a simpler and more practical alternative to the systematic nomenclature. It has developed gradually over time and is only applicable to alkanes with basic structures. It can only be used for alkanes with relatively simple structures, and systematic nomenclature must be used for more complex alkanes.

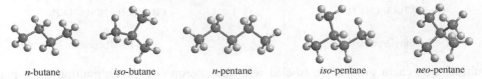

n-butane　　　*iso*-butane　　　*n*-pentane　　　*iso*-pentane　　　*neo*-pentane

The parent name of a **cyclic structure** is preceded by the primary prefix cyclo-. For example, according to IUPAC conventions, a six-carbon alkane ring is a cyclohexane.

6.1.2　Alkenes

Alkenes and cycloalkenes are hydrocarbons containing one or more C=C double bonds. The compounds are said to be unsaturated because they do not have the maximum number of atoms that each carbon can accommodate. The parent name of an alkene is the stem name with the suffix -ene. The IUPAC nomenclature for alkenes follows rules similar to those for alkanes. The longest continuous sequence of carbon atoms containing the double bond is the basis for the parent name of an alkene. The location of the double bond is determined by numbering the parent chain from the end that provides the smallest number to the double bond.

$$\underset{1\ \ 2\ \ \ \ \ 3\ \ 4}{CH_3CH=CHCH_3} \qquad \underset{1\ \ \ \ \ 2\ \ \ 3\ \ 4}{H_2C=CHCH_2CH_3} \qquad \underset{1\ \ \ \ \ 2\ \ \ 3\ \ 4\ \ 5\ \ 6}{H_2C=CHCHCH_2CH_2CH_3}\overset{CH_2CH_2CH_3}{|}$$

2-butene　　　　　　1-butene (NOT 3-butene)　　　3-propyl-1-hexene (NOT name as a heptane)

The location of the double bond in an alkene is the determining factor for numbering the atoms in the IUPAC system, even if the side chain alkyl groups have higher numbers. For instance, 5-methyl-3-heptene is not referred to as 3-methyl-4-heptene.

$$\overset{CH_3}{\underset{}{|}}\ H_3CC=CHCH_2CH_3 \qquad\qquad CH_3CH_2CH=CHCHCH_2CH_3\overset{CH_3}{\underset{}{|}}$$

2-methyl-2-pentene　　　　　　　　5-methyl-3-heptene (NOT 3-methyl-4-heptene)

When naming a **cycloalkene** compound, the numbering starts at one carbon atom of the double bond, moves to the second atom of the double bond and then around the ring. Within this constraint, the ring is numbered so that the side chains have the lowest possible numbers. The number -1 is not required to indicate the position of the double bond. If a side chain features a C=C double bond, its substituent name ends in -enyl. The "ene" of the parent name for the side chain is retained to indicate a double bond, but the last "e" is replaced with "-yl" to represent a substituent group. The side chain is numbered from the point of attachment to the parent chain.

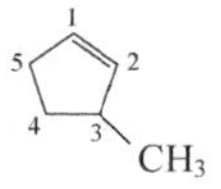

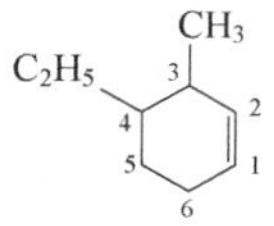

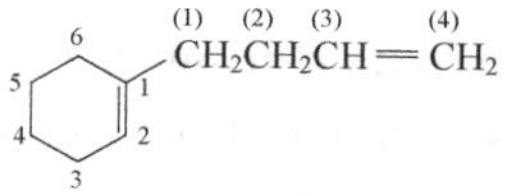

3-methylcyclopentene (NOT 2-methylcyclopentene)

4-ethyl-3-methylcyclohexene (NOT 3-ethyl-2-methylcyclohexene)

1-(3-butenyl)cyclohexene

The following three unsaturated side chain groups may retain their common names in the IUPAC nomenclature:

$CH_2=CH-$ vinyl

$CH_2=CHCH_2-$ allyl

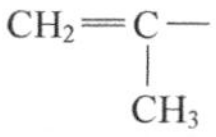

$CH_2=C(CH_3)-$ isopropenyl

Many organic compounds contain two or more double bonds. Doubly unsaturated hydrocarbons are called **dienes**, hydrocarbons with three double bonds are called **trienes**, and so on. Unsaturated hydrocarbons with more than one double bond are called polyenes. Dienes in which two double bonds are separated by only one single bond are known as conjugated dienes (>C=C–C=C<). They have chemical properties that distinguish them from simple alkenes. 1,3-butadiene (commonly known as butadiene) is a prominent member of this group and is the starting material for some polymers. Dienes and polyenes in which the double bonds are separated by more than one single bond are non-conjugated and are chemically similar to simple alkenes.

$CH_2=CH-CH=CH_2$
1, 3-butadiene

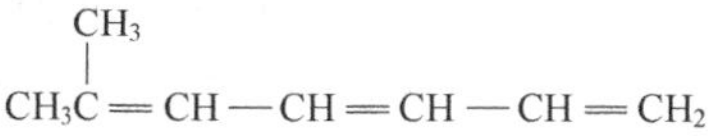

$CH_3C(CH_3)=CH-CH=CH-CH=CH_2$
6-methyl-1, 3, 5-heptatriene (conjugated)

$CH_2=CH-CH_2-CH=CH_2$
1, 4-pentadiene (non-conjugated)

Compounds in which a single carbon atom is linked to two other carbon atoms by double bonds (>C=C=C<) are called **allenes**. Allenes are often difficult to prepare and are relatively reactive.

Aromatic hydrocarbons are a class of polyenes that are very different from the common alkenes. They are generally referred to as arenes. Most arenes are related to the six-membered conjugated carbocycle benzene. Their chemical properties are associated with a unique type of conjugation known as aromaticity. Aromatic compounds are named as derivatives of benzene or related parent structures. When two substituents are attached to the benzene ring, their positions are indicated by numbering or, more commonly, by the prefixes *ortho-* (*o*), *meta-* (*m*), or *para-* (*p*). The prefix *ortho-* means that the substituents are in a 1, 2-ratio, *meta-* represents a 1, 3-ratio, and *para-* a 1, 4-ratio.

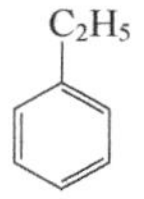

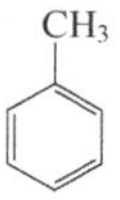

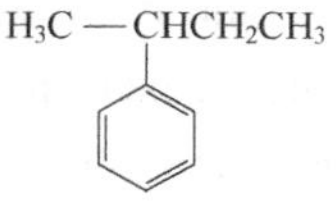

benzene

ethylbenzene

methylbenzene (toluene)

2-butylbenzene

The terms *ortho*, *meta*, and *para* are prefixes used in organic chemistry to indicate the position of non-hydrogen substituents on a hydrocarbon ring (benzene derivative). The prefixes are derived from Greek words meaning correct/straight, following/after, and the like. *Ortho*-, *meta*-, and *para*- have historically had different meanings, but in 1879 the American Chemical Society agreed on the following definitions, which are still in use today. The positions of the double bonds in the aromatic ring are not taken into account in the numbering.

1, 2-dimethylbenzene   1-ethyl-3-methylbenzene   1, 4-diethylbenzene
(*o*-dimethylbenzene)   (*m*-ethylmethylbenzene)   (*p*-diethylbenzene)

The substituent group derived from benzene by the conceptual removal of a hydrogen atom (C_6H_5-) is called phenyl, not benzyl; the benzyl substituent is $C_6H_5CH_2-$.

1, 3-diphenylpropane   triphenylmethane

6.1.3 Alkynes

Alkynes are unsaturated hydrocarbons containing one or more carbon-carbon triple bonds. The simplest alkyne is the important industrial gas ethyne (acetylene, C_2H_2). Alkynes are often referred to as acetylenes, a non-IUPAC name. The longest continuous carbon chain containing the triple bond is the basis for the parent name of an alkyne. The suffix -yne is added to the stem.

$CH_3C \equiv CCH_3$ $(CH_3)_2CHCH_2C \equiv CH$
2-butyne   4-methyl-1-pentyne

If there are both double and triple bonds in the parent chain, the ending becomes -enyne. The numbering is such that the double and triple bonds are given the lowest possible numbers, whether -ene or -yen gets the lower number. If both groups could be given the same number, -ene takes precedence and is given the lower number.

6.2 Heteroatom Functional Groups

In the previous section, we learned that the double and/or triple bonds of alkenes and alkynes are often the positions at which chemical reactions take place. However, the majority of functional groups found in organic molecules involve carbon attached to various heteroatoms. A wide variety of chemical reactions can take place on the carbon or heteroatoms of these groups.

The remainder of this unit introduces the important classes of compounds containing heteroatomic functional groups that are important in organic chemistry. It will also cover a brief nomenclature of these organic compounds containing heteroatoms.

6.2.1 Alcohols, Phenol, and Thiols

Alcohols are probably the organic compounds with which people are most familiar. The functional part of an alcohol is a hydroxy group attached to an alkyl group. An alcohol can be thought of as a derivative of water in which a hydrogen atom is replaced by the carbon atom of an organic molecule. Many properties of low molecular weight alcohols are similar to those of water.

$$H-O-H \qquad \qquad -\overset{|}{\underset{|}{C}}-O-H$$

water alcohol

Alcohols are designated by the suffix -ol in the IUPAC nomenclature. The longest continuous chain to which the hydroxy group is attached forms the root of the parent name. The last "e" of the corresponding hydrocarbon name is dropped and the suffix "-ol" is added. Numbering again begins at one end of the parent carbon chain, and that end is chosen to give the lowest possible number to the hydroxy group.

CH_3OH CH_3CH_2OH $CH_3CH_2CH_2CH_2OH$ $CH_3\overset{\overset{OH}{|}}{CH}CH_2CH_2CH_3$ 3-methylcyclopentanol

methanol   ethanol   1-butanol   2-pentanol   3-methylcyclopentanol

Compounds containing two, three or more hydroxy groups are classified as polyols. The IUPAC suffixes are -diol, -triol, etc. Glycol is the common name for 1,2-diols. 1,2-ethanediol is an important industrial product used as an antifreeze. The IUPAC name retains the final "e" of the parent hydrocarbon name because the suffix "-diol" does not begin with a vowel. We also find some examples of nomenclature where a letter is added or deleted to make pronunciation easier.

$$HO-CH_2CH_2-OH$$

1, 2-ethanediol (ethylene glycol)

When a compound contains **more than one functional group**, a choice must be made as to which group to use in the parent name. The IUPAC nomenclature gives a priority order for functional groups.

For example, alcohols take precedence over alkenes. In an alkenol, a compound containing a C=C double bond and a hydroxy group, the parent name and numbering are chosen so that the hydroxy group has the lowest number, whether or not the double bond is part of the parent chain[2]. If another functional group takes precedence over the alcohol in the molecule, the hydroxy group is called the hydroxy substituent.

$CH_2=CHCH_2OH$ $(CH_3)_2CH\overset{\overset{OH}{|}}{C}H\overset{}{C}HCH_2CH_2CH_3$
                                                                        $CH=CH_2$

2-propen-1-ol(allyl alcohol)   4-ethylene-2-methyl-3-heptanol

A common division of alcohols is based on the number of carbon atoms attached to the hydroxy group. For example, 1-propanol is a primary alcohol because there is only one carbon atom attached to the hydroxy group, while 2-propanol is a secondary alcohol and 2-methyl-2-propanol is a tertiary alcohol.

$CH_3CH_2CH_2OH$　　　　　CH_3CHCH_3 with OH　　　　　$H_3C-C(OH)(CH_3)-CH_3$

1-propanol　　　　2-propanol (isopropyl alcohol)　　　　2-methyl-2-propanol (*tert*-butyl alcohol)
a primary alcohol　　　　a secondary alcohol　　　　a tertiary alcohol

The hydroxy group is found in a wide variety of compounds of plant and animal origin. Because of the complex structures of some of these substances, they are usually referred to by **common names**.

menthol (from oil of peppermint)　　　　$(CH_3)_2C=CHCH_2CH_2C(CH_3)=CHCH_2OH$

geraniol (from rose oil and geraniums)

When the hydroxy group is attached to a carbon atom of a benzene ring, the compound is called a **phenol**. The simplest member of the class, hydroxybenzene, is called a phenol. Phenols are often considered separately from alcohols because their chemical properties are quite different. Common names are used for many phenols.

hydroxybenzene　　　　1, 2-dihydroxybenzene　　　　1, 3-dihydroxybenzene
(phenol, m.p. 42℃)　　　　(catechol, m.p. 103℃)　　　　(resorcinol, m.p. 111℃)

1, 4-dihydroxybenzene　　　　1-hydroxy-3-methylbenzene　　　　1-hydroxy-4-methylbenzene
(hydroquinone, m.p. 173℃)　　　　(*m*-cresol, m.p. 10℃)　　　　(*p*-cresol, m.p. 32℃)

Thiols are the sulfur analogues of alcohols, in which a sulfur atom replaces the oxygen atom in the functional group. They are named by adding the suffix -thiol to the parent name. Thiols are also known as **mercaptans**.

CH_3SH　　　　　　　　　　　　　　　CH_3CH_2SH

methanethiol (methyl mercaptan, b.p. 6℃)　　　　ethanethiol (ethyl mercaptan, b.p. 35℃)

6.2.2 Ethers

Ethers are compounds in which two carbon atoms are bonded to a single oxygen atom. The nomenclature of ether is somewhat confusing, as no system has been officially adopted by IUPAC.

The current preferred practice is to refer to one of the alkyl groups plus the oxygen atom (RO–) as an alkoxy. This alkoxy is then treated as a substituent on a parent structure of the rest of the molecule. Most simple ethers are named dialkyl or alkyl alkyl ethers, e.g. $CH_3CH_2OCH_2CH_3$ is diethyl ether and $CH_3CH_2OCH_3$ is ethyl methyl ether.

$CH_3OCH(CH_3)_2$

2-methoxypropane (isopropyl methyl ether)

▷—OCH_2CH_3

ethoxycyclopropane (cyclopropyl ethyl ether)

Another approach, which seems to be the closest to systematic nomenclature, is to treat the oxygen atom as if it were an atom of the parent chain. The oxygen atom is counted as a carbon in determining the parent name and is then designated by the prefix -oxa with a number indicating its position in the parent chain.

$$CH_3CH_2OCH_2CH_2CH_3$$
$$1 \quad 2 \quad 3\ 4 \quad 5 \quad 6$$

3-oxahexane (ethyl propyl ether)

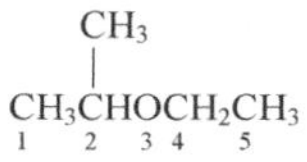

$$CH_3\overset{\underset{|}{CH_3}}{CH}OCH_2CH_3$$
$$1 \quad 2 \quad 3 \quad 4 \quad 5$$

2-methyl-3-oxapentane (ethyl isopropyl ether)

When oxygen is an atom in a three-membered ring, the systematic name for the **cyclic ether** is oxirane or oxacyclopropane. Compounds such as oxirane are commonly called epoxides and are named as the oxidation products of the corresponding alkenes.

$$CH_2\!=\!CH_2 \xrightarrow[\text{Ag catalyst}]{O_2}$$

ethene (ethylene)

$$\overset{O}{\underset{H_2C—CH_2}{\triangle}}$$

oxirane or oxacyclopropane (ethylene oxide)

Compounds in which an atom other than carbon is part of a ring are called **heterocycles**. In the case of cyclic ethers, the heteroatom is oxygen. We will see many examples of oxygen, sulfur, and nitrogen heterocycles throughout our study of organic compounds.

oxacyclopentane
(tetrahydrofuran, THF)

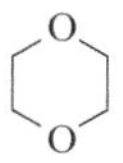

1, 4-dioxacyclohexane
(1,4-dioxane)

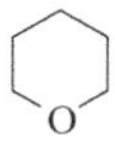

oxacyclohexane
(tetrahydropyran)

6.2.3 Amines

Amines are organic derivatives of ammonia (NH_3). They are the most important of the compounds in organic chemistry that behave as bases. One, two or three alkyl groups can replace the hydrogens of ammonia to form primary, secondary, or tertiary amines. Note that the terms "primary" "secondary", and "tertiary" have a related structural meaning for amines and for alcohol. For amines, these terms refer to the number of carbon atoms attached to the nitrogen atom, and for alcohols, they refer to the number of carbon atoms on the hydroxy-bearing carbon (the carbinol carbon atom).

CH_3NH_2 $(CH_3)_2NH$ $(CH_3)_3N$
methanamine   N-methylmethanamine   N,N-dimethylmethanamine
(methylamine)   (dimethylamine)   (trimethylamine)
a primary amine   a secondary amine   a tertiary amine

In the IUPAC nomenclature for primary amines, the last "e" of the parent hydrocarbon name is dropped and the suffix -amine is added. The nitrogen atom is neither counted nor numbered in determining the parent name. The alkyl groups of secondary and tertiary amines are named as substituents to the parent and the letter "N" is used to indicate that the substituent is attached to the nitrogen atom. In these cases, the parent name is based on the largest or most complex alkyl group. It is also common to use the hydrocarbon parent name as if it were a substituent with the suffix -ylamine. This corresponds to naming amines as hydrocarbon derivatives of ammonia.

$(CH_3)_2CHNH_2$ $(CH_3CH_2)_2NH$
2-propanamine   N-ethylethanamine
(isopropylamine)   (diethylamine)

⬡—$N(CH_3)_2$ $(CH_3)_2NCH_2CH_2C(CH_3)_3$

N,N-dimethylcyclohexanamine   3,3,N,N-tetramethylbutanamine

When four carbon atoms are attached to the nitrogen, the compound is no longer basic. The tetracoordinate nitrogen atom has a positive charge and is the cationic part of a type of compound known as a **quaternary ammonium salt**.

$(CH_3)_4N^+I^-$

tetramethylammonium iodide a quaternary ammonium salt

The closely related ammonium salts are formed by the addition of a proton (H^+) to ammonia or amines in a rapid acid-base reaction.

NH_3 +   HCl   ⇌ $NH_4^+Cl^-$
ammonia   ammonium chloride

$(CH_3)_2NH$ +   HCl   ⇌ $(CH_3)_2NH_2^+Cl^-$
N-methylmethanamine (dimethylamine)   dimethylammonium chloride

The prefix amino- or alkylamino- is used when the amine is a substituent on a parent molecule. In some cases, a parent chain containing a nitrogen atom is identified by the addition of the prefix -aza, analogous to the use of -oxa in the naming of ethers.

2-(N,N-dimethylamino)cyclopentanol   azacyclohexane (piperidine)

Part B Introduction to Fundamentals of Chemistry and Chemical Engineering   097

Anilines are amines in which the amino group is attached to a carbon atom of the benzene ring.

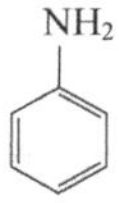

C₆H₅NH₂ — benzenamine (aniline)

C₆H₅NHCH₂CH₃ — N-ethylbenzenamine (N-ethylaniline)

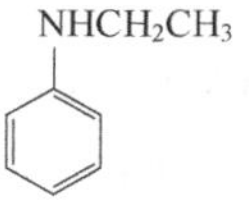

Compounds containing one or more nitrogen atoms are among the most abundant organic substances in nature. **Alkaloids** are nitrogenous bases (often heterocyclic) that are widely distributed in plants and often have important physiological activities. **Amino acids** are the building blocks of proteins.

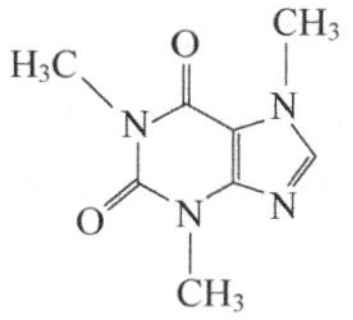

caffeine 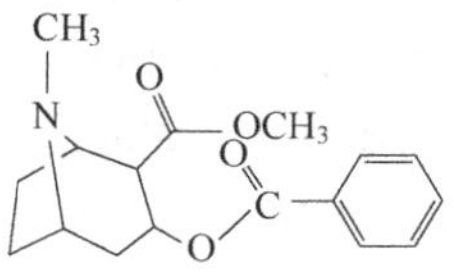   cocaine   glycine $H_2NCH_2CO_2H$ tryptophan

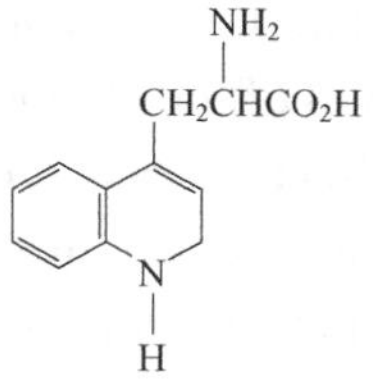

plant alkaloids   amino acids

Putrescine and cadaverine are two particularly pungent diamines produced by the bacterial decomposition of animal matter.

$H_2NCH_2CH_2CH_2CH_2NH_2$

1, 4-butanediamine (putrescine)

$H_2NCH_2CH_2CH_2CH_2CH_2NH_2$

1, 5-pentanediamine (cadaverine)

6.2.4 Organohalogen Compounds

Substances in which one or more halogen atoms are bonded to carbon are known as organohalogen compounds. Although they are often referred to and named as organic halides, these materials are not ionic as this term might imply. Organohalogen compounds are called substituted hydrocarbons because there is no suffix to represent halogen.

CH_3CH_2F

fluoroethane

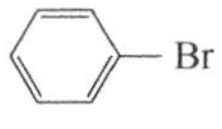

C₆H₅—Br

bromobenzene

$CH_3\underset{|}{\overset{Br}{C}}HCH=CHCH_3$

4-bromo-2-pentene

CH_3Cl

chloromethane
(methyl chloride)

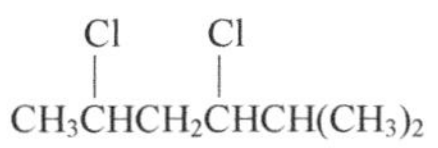

$CH_3\underset{|}{\overset{Cl}{C}}HCH_2\underset{|}{\overset{Cl}{C}}HCH(CH_3)_2$

2, 4-dichloro-5-methylhexane
(NOT 3, 5-dichloro-2-methylhexane)

$(CH_3)_2\underset{|}{\overset{I}{C}}CH_2CH_2C\equiv CH$

5-iodo-5-methyl-1-hexyne*

*Double and triple bonds and substituents other than saturated hydrocarbons take precedence over halogen assignment of numbers.

6.2.5 Aldehydes and Ketones

Aldehydes and ketones are often called carbonyl compounds because they contain carbonyl groups >C=O. The carbonyl group is chemically one of the most versatile functional groups we will encounter. We will see that a large number of important reactions involve modifications to carbonyl groups.

$$\underset{\text{an aldehyde}}{R-\overset{\overset{O}{\|}}{C}-H} \qquad \underset{\text{a ketone}}{R-\overset{\overset{O}{\|}}{C}-R \quad \text{or} \quad R-\overset{\overset{O}{\|}}{C}-R'}$$

(R and R' represent alkyl or alkyl groups)

The IUPAC nomenclature uses the suffix -one to denote a ketone. A number indicates the position of the carbonyl group along the parent chain. It is also common to name ketones using the substituent names for the two hydrocarbon groups attached to the carbonyl, followed by the word ketone, and the three words should be written separately. When another functional group has nomenclature priority over ketone, the oxygen atom of the ketone carbonyl group is considered a substituent and is designated by the prefix -oxo. Note that the closely related suffixes -oxa and -aza are used for ethers and amines, respectively.

$$\underset{\text{3-pentanone (diethyl ketone)}}{CH_3CH_2\overset{\overset{O}{\|}}{C}CH_2CH_3} \qquad \underset{\text{3-methyl-2-butanone (isopropyl methyl ketone)}}{CH_3\overset{\overset{O}{\|}}{C}CH(CH_3)_2}$$

3-chlorocyclohexanone

$$\underset{\text{2, 4-pentanedione}}{CH_3\overset{\overset{O}{\|}}{C}CH_2\overset{\overset{O}{\|}}{C}CH_3}$$

Aldehydes are named with the suffix -al. When a compound is named an aldehyde, the functional group is necessarily at the end of the parent chain, so the number 1- is omitted. The suffix -carbaldehyde or -carboxaldehyde is used when the aldehyde group (–CHO) is attached to a ring. When the aldehyde group is a substituent on a parent of higher nomenclature priority, the term methanoyl- or formyl- is used.

$$\underset{\text{propanal (propionaldehyde)}}{CH_3CH_2\overset{\overset{O}{\|}}{C}-H} \qquad \underset{\text{4-oxopentanal}}{CH_3\overset{\overset{O}{\|}}{C}CH_2CH_2\overset{\overset{O}{\|}}{C}-H}$$

$$\underset{\text{2-methylpropanal}}{(CH_3)_2CH\overset{\overset{O}{\|}}{C}-H} \qquad \underset{\text{benzencarbaldehyde (benzaldehyde)}}{C_6H_5-\overset{\overset{O}{\|}}{C}-H}$$

6.2.6 Carboxylic Acid

The most important acids in organic chemistry belong to the class of compounds known as carboxylic acids. Carboxylic acids were one of the earliest types of organic compounds studied by ancient chemists and are found in or derived from many natural substances. The names commonly used for carboxylic acids derive from the early sources of these compounds.

$$R-\overset{\overset{O}{\|}}{C}-OH$$

a carboxylic acid

The suffix -oic acid is used by IUPAC to denote carboxylic acids. The carboxylic acid group must be terminal, so the number 1- is not included. When the functional group is attached to a cyclic structure, -carboxylic acid becomes the appropriate suffix. The term -carboxy is used when the carboxylic acid functional group is designated as a substituent.

benzenecarboxylic acid (benzoic acid)

$(CH_3)_2CHCH_2CH_2\overset{\overset{O}{\|}}{C}-OH$

4-methylpentanoic acid

$(CH_3)_3C\overset{\overset{O}{\|}}{C}CH_2\overset{\overset{O}{\|}}{C}CH_2-OH$

5,5-dimethyl-3-oxohexanoic acid

$CH_3CH=CHCH_2\overset{\overset{NH_2}{|}}{CH}CH_2\overset{\overset{O}{\|}}{C}-OH$

3-amino-5-heptenoic acid

The older names for carboxylic acids — formic rather than methanoic, acetic rather than ethanoic, etc. — are still commonly used for compounds with five or fewer carbon atoms. However, there is a tendency to phase these out in favor of the IUPAC systematic nomenclature. An older system for indicating the position of substituents in carboxylic acids uses Greek letters rather than the IUPAC numbering scheme. The carbon atom adjacent to the carboxy group is called the alpha (α) carbon, the next beta (β), etc., rather than the IUPAC system where the carboxy carbon is numbered 1. The Greek letter system is discouraged, but it is encountered often enough that we should be familiar with it.

$CH_3CH_2CH_2\overset{\overset{Cl}{|}}{CH}CH_2\overset{\overset{O}{\|}}{C}-OH$

3-chlorohexanoic acid

(β-chlorohexanoic acid)

$CH_3CH_2CH_2\overset{\overset{O}{\|}}{\underset{\underset{NH_2}{|}}{CH}}\overset{}{C}-OH$

2-aminopentanoic acid

(α-aminopentanoic acid)

6.2.7 Carboxylic Acid Derivatives

Compounds in which the hydroxy group of a carboxylic acid is replaced by another heteroatom or group are known as carboxylic acid derivatives. The formation of carboxylic acid derivatives involves the replacement of hydroxy with halogen (**acyl halides**), carboxylate (**carboxylic acid anhydrides**), alkoxy (**esters**), and amino (**amides**) groups. The derivatives are readily interconverted and can be hydrolyzed to the parent carboxylic acids. Aldehydes and ketones are not carboxylic acid derivatives.

$$\begin{array}{ccccc} \overset{O}{\underset{\|}{-C}}-OH & \overset{O}{\underset{\|}{-C}}-X & \overset{O}{\underset{\|}{-C}}-O-\overset{O}{\underset{\|}{C}}- & \overset{O}{\underset{\|}{-C}}-OR & \overset{O}{\underset{\|}{-C}}-N\diagup \\ \text{carboxylic acid} & \text{acyl halide} & \text{carboxylic anhydride} & \text{ester} & \text{amide} \end{array}$$

Acyl halides are named by adding the suffix -oyl to the parent name followed by a separate word to designate the specific halogen atom. When common acid names are used as the parent, -yl is the suffix.

$$\begin{array}{ccc} H_3C-\overset{O}{\underset{\|}{C}}-Cl & C_2H_5-\overset{O}{\underset{\|}{C}}-Cl & C_6H_5-\overset{O}{\underset{\|}{C}}-Cl \\ \text{ethanoyl chloride} & \text{propanoyl chloride} & \text{benzoyl chloride} \\ \text{(acetyl chloride, b.p. 51℃)} & \text{(propionyl chloride b.p. 80℃)} & \text{(b.p. 197℃)} \end{array}$$

Carboxylic acid anhydrides are named after the corresponding carboxylic acid followed by the word anhydride. If both groups attached to the common oxygen atom are the same, the acid name is used without the prefix di-. If two different groups are attached to the oxygen atom, their carboxylic acid names are used as separate words.

cyclopropanecarboxylic propanoic anhydride   butanedioic anhydride (succinic anhydride)

Carboxylic acid esters are named using a combination of the names of the alcohol and carboxylic acid components of the molecule. The parent alcohol name is used with a -yl suffix as if it were a substituent, but written as a separate first word. The second word is the parent name for the carboxylic acid, with the -oate suffix replacing -oic acid. If the substituent is an ester group, use -alkoxycarbonyl or -carboalkoxy is used. Salts of carboxylic acids have a similar nomenclature, with the cation name given first.

ethyl cyclohexanecarboxylate   sodium benzoate   phenyl ethanoate   cyclohexyl propanoate
                                                   (phenyl acetate)   (cyclohexyl propionate)

$$\begin{array}{ccc} H_3C-\overset{O}{\underset{\|}{C}}-NH_2 & CH_3CH=CHC-NH_2 & H_3CH_2C-\overset{O}{\underset{\|}{C}}-N(C_2H_5)_2 \\ \text{ethanamide} & \text{2-butenamide} & N,N\text{-diethylpropanamide} \\ \text{(acetamide, m.p. 82℃)} & \text{(crotonamide, m.p. 162℃)} & \text{(b.p. 191℃)} \end{array}$$

Nitriles are often considered together with carboxylic acid derivatives because the cyano group $-C\equiv N$ is easily converted to the carboxylic acid group. Their nomenclature reflects their relationship to carboxylic acids. The suffix -nitrile or -onitrile is added to their parent name and the numbering includes the nitrile carbon atom. The prefix cyano- is used to denote the nitrile group as a substituent.

$$\underset{\text{3-chlorobutanenitrile}}{CH_3\overset{Cl}{\underset{|}{C}}HCH_2C\equiv N}\qquad \underset{\text{hexanedinitrile (adiponitrile)}}{N\equiv CCH_2CH_2CH_2CH_2C\equiv N}\qquad \underset{\text{3-cyanobutanoic acid}}{CH_3\overset{C\equiv N}{\underset{|}{C}}HCH_2COOH}$$

$$\underset{\text{ethanenitrile (acetonitrile)}}{H_3CC\equiv N}\qquad\qquad \underset{\text{propenenitrile (acrylonitrile)}}{H_2C=CHC\equiv N}$$

New Words and Expressions

hydrocarbon	[ˌhaɪdrəˈkɑːbən]	n. 碳氢化合物
aliphatic	[ˌælɪˈfætɪk]	adj. 脂肪族的
alicyclic	[ˌælɪˈsaɪklɪk]	adj. 脂环族的，脂环的
acyclic	[ˌeɪˈsaɪklɪk]	adj. 无环的，非周期的
property	[ˈprɒpəti]	n. 属性
paraffin	[ˈpærəfɪn]	n. 链烷烃
methylene	[ˈmeθɪˌliːn]	n. 亚甲基
butane	[ˈbjuːteɪn]	n. 丁烷
prefix	[ˈpriːfɪks]	n. 前缀
nomenclature	[nəˈmeŋklətʃə(r)]	n. 命名法，术语
fragment	[ˈfrægmənt]	n. 碎片，片段
hyphen	[ˈhaɪfn]	n. 连字符，连字号
designate	[ˈdezɪgneɪt]	v. 把……描述为
comma	[ˈkɒmə]	n. 逗号
flexible	[ˈfleksəb(ə)l]	adj. 灵活的，柔韧的
cyclic	[ˈsaɪklɪk]	adj. 环的，循环的
cyclohexane	[ˌsaɪkləʊˈheksein]	n. 环己烷
cycloalkene	[saɪkˈləʊlkiːn]	n. 环烯
criterion	[kraɪˈtɪəriən]	n. 尺度，标准，准则
constraint	[kənˈstreɪnt]	n. 限制，束缚，克制
diene	[ˈdaɪiːn]	n. 二烯
polyene	[ˈpɒlijiːn]	n. 多烯
conjugate	[ˈkɒndʒəgeɪt]	n. 结合物，共轭物
polymer	[ˈpɒlɪmə(r)]	n. 聚合物
allene	[ˈæliːn]	n. 丙二烯
arene	[ˈæriːn]	n. 芳烃
conceptual	[kənˈseptʃuəl]	adj. 概念的，观念的
phenyl	[ˈfiːnaɪl]	n. 苯基

benzyl	[ˈbenzaɪl]	n.	苄基，苯甲基
acetylene	[əˈsetəliːn]	n.	乙炔
suffix	[ˈsʌfɪks]	n.	后缀，词尾
previous	[ˈpriːviəs]	adj.	以前的，先前的
briefly	[ˈbriːfli]	adv.	简短地，简要地
corresponding	[ˌkɒrəˈspɒndɪŋ]	adj.	相应的，相关的
hydroxy	[haɪˈdrɒksi]	adj.	氢氧根的
polyol	[ˈpɒlɪˌɒl]	n.	多元醇
antifreeze	[ˈæntifriːz]	n.	防冻剂
glycol	[ˈglaɪkɒl]	n.	乙二醇，甘醇
vowel	[ˈvaʊəl]	n.	元音
pronunciation	[prəˌnʌnsiˈeɪʃ(ə)n]	n.	发音方法，读法
propanol	[ˈprəʊpənɒl]	n.	丙醇
peppermint	[ˈpepəmɪnt]	n.	薄荷，薄荷油
geranium	[dʒəˈreɪniəm]	n.	天竺葵
phenol	[ˈfiːnɒl]	n.	苯酚
thiol	[ˈθaɪəʊl]	n.	硫醇
sulphur	[ˈsʌlfə(r)]	n.	硫，硫黄
analogue	[ˈænəlɒg]	n.	相似物
alkoxy	[ælˈkɒksi]	n.	烷氧基
oxacyclopropane	[ɒksəaɪkləʊˈprəʊpeɪn]	n.	氧杂环丙烷，环氧乙烷
epoxide	[ɪˈpɒksaɪd]	n.	环氧化物
oxidation	[ɒksɪˈdeɪʃn]	n.	氧化
amine	[əˈmiːn]	n.	胺
isopropylamine	[aɪsəprɒpɪˈlæmaɪn]	n.	异丙胺
diethylamine	[daɪeθɪləˌmiːn]	n.	二乙胺
tetracoordinate		adj.	四配位的
cationic	[ˌkætaɪˈɒnɪk]	adj.	阳离子的
ammonium	[əˈməʊniəm]	n.	铵
alkylamino	[ɔːlkɪˈlæmɪnəʊ]	n.	烷氨基
aniline	[ˈænɪlɪn]	n.	苯胺；adj. 苯胺的
alkaloid	[ˈælkəlɔɪd]	n.	生物碱，植物碱基
caffeine	[ˈkæfiːn]	n.	咖啡因
cocaine	[kəʊˈkeɪn]	n.	可卡因
glycine	[ˈglaɪsiːn]	n.	甘氨酸
tryptophan	[ˈtrɪptəˌfæn]	n.	色氨酸
putrescine	[pjuːˈtresiːn]	n.	腐胺
cadaverine	[kəˈdævəˌriːn]	n.	尸胺，1, 5-戊二胺
diamine	[ˈdaɪəmiːn]	n.	二元胺
bacterial	[bækˈtɪəriəl]	adj.	细菌的
decomposition	[ˌdiːˌkɒmpəˈzɪʃn]	n.	分解，腐烂，变质
versatile	[ˈvɜːsətaɪl]	adj.	多用途的，多功能的

priority	[praɪˈɒrəti]	n. 优先事项，优先
carbaldehyde		n. 甲醛
methanoyl	[ˈmeθənɔɪl]	n. 甲酰氧基
terminal	[ˈtɜːmɪnl]	n. 终点站；adj. 致命的
methanoic		adj. 甲酸的
acetic	[əˈsiːtɪk]	adj. 醋酸的
ethanoic	[ˌeθəˈnəʊɪk]	adj. 乙酸的
scheme	[skiːm]	n. 计划，方案
chlorohexanoic		adj. 氯己酸的
aminopentanoic		adj. 氨基戊酸的
anhydride	[ænˈhaɪdraɪd]	n. 酸酐
succinic	[sʌkˈsɪnɪk]	adj. 琥珀的
component	[kəmˈpəʊnənt]	n. 成分，部件
alkoxycarbonyl	[ɔːlkɒksiːˈkɑːbənɪl]	n. 烷氧羰基
ethanamide	[ɪˈθænəmaɪd]	n. 乙酰胺
butenamide		n. 丁烯酰胺
diethylpropanamide		n. 二乙基丙酰胺
nitrile	[ˈnaɪtrɪl]	n. 腈
adiponitrile	[ædɪpəʊˈnaɪtriːl]	n. 己二腈
acetonitrile	[əˌsiːtəʊˈnaɪtrɪl]	n. 乙腈
acrylonitrile	[ˌækrɪlə(ʊ)ˈnaɪtraɪl]	n. 丙烯腈

Notes

1. For compounds more complex than normal hydrocarbons, the name of the normal hydrocarbon derived from the longest chain of carbon atoms in the compound is the parent name upon which further nomenclature is based.
   对于比直链烃更复杂的化合物而言，进一步对其命名的依据是把从化合物中最长的碳原子链衍生出的直链烃的名称作为母体名称。

2. In an alkenol, a compound containing a C=C double bond and a hydroxy group, the parent name and numbering are chosen so that the hydroxy group has the lowest number, whether or not the double bond is part of the parent chain.
   烯醇为一种含有碳碳双键和羟基的化合物，无论其双键是否是母链的一部分，母体名称和编号的选择都要使羟基的编号最小。

Trivial — Last but not the Least

Affixes and Abbreviations
(1) Organic chemistry prefixes and suffixes

The purpose of organic chemistry nomenclature is to indicate how many carbon atoms are in a chain, how the atoms are bonded together, and the identity and location of any functional groups in the molecule. The root names of hydrocarbon molecules are based on whether they form a chain or a ring. The name is preceded by a prefix. The prefix to the name of the molecule is based on the number of carbon atoms. For example, a chain of six carbon atoms would be named with the prefix hex-. The suffix to the name is an ending that describes the types of chemical bonds in the molecule. An IUPAC name also includes the names of the substituent groups (other than hydrogen) that make up the molecular structure.

(2) Hydrocarbon suffixes

The suffix or ending of the name of a hydrocarbon depends on the nature of the chemical bonds between the carbon atoms. The suffix is -ane if all the carbon-carbon bonds are single bonds (formula C_nH_{2n+2}), -ene if at least one carbon-carbon bond is a double bond (formula C_nH_{2n}), and -yne if there is at least one carbon-carbon triple bond (formula C_nH_{2n-2}). There are other important organic suffixes:

-ol means that the molecule is an alcohol or contains the —C—OH functional group.

-al means that the molecule is an aldehyde or contains the O=C—H functional group.

-amine means that the molecule is an amine with the —C—NH_2 functional group.

-ic acid indicates a carboxylic acid containing the O=C—OH functional group.

-ether indicates an ether having the —C—O—C— functional group.

-ate is an ester having the functional group O=C—O—C.

-one is a ketone having the —C=O functional group.

(3) Hydrocarbon prefixes

Table 6.1 in this unit lists the organic chemical prefixes, or the root words, for up to 20 carbons in a simple hydrocarbon chain. It would be a good idea to memorize this table early in your organic chemistry studies.

(4) Common names

Note that hydrocarbons found as rings (aromatic hydrocarbons) have slightly different names. For example, C_6H_6 is called benzene. Because it contains carbon-carbon double bonds, it has the suffix -ene. However, the prefix actually comes from the word "gum benzoin", which has been used as an aromatic resin since the 15th century. When the hydrocarbons are substituents, there are several common names you may come across:

amyl: substituent with 5 carbons

valeryl: substituent with 5 carbons

lauryl: substituent with 12 carbons

myristyl: substituent with 14 carbons

cetyl or palmityl: substituent with 16 carbon atoms

stearyl: substituent with 18 carbon atoms

phenyl: common name for a hydrocarbon with benzene as a substituent

(5) Functional group priorities and abbreviations

When two or more functional groups are part of the same molecule, a choice must be made as to which will be the basis of the parent name, and the suffix, root name and numbering will depend on this decision.

An order of priority has been established for the IUPAC nomenclature. The highest priority is given to functional groups that have an IUPAC suffix and that must end a carbon chain. Carboxylic acids and their derivatives are in this category. Next are groups that have a suffix and can be located

Part B Introduction to Fundamentals of Chemistry and Chemical Engineering   105

anywhere in a molecule. Examples are hydroxy (alcohol) and amino (amine). The lowest priority is given to groups such as halogen, which have no suffix and are therefore always called substituents. With the exception of -ene or -yne for double and triple bonds, which may become part of the parent name, only one suffix is used with the parent name in the IUPAC system. Table 6.2 below lists some groups that are not considered in this unit, but are important for later study.

Table 6.2 Summary of IUPAC nomenclature arranged in order of decreasing numerical priority

Class	Formula	Suffix	Prefix
cations	R_4N^+	-ammonium	ammonio-
	R_4P^+	-phosphonium	phosphonio-
	R_3S^+	-sulfonium	sulfonio-
carboxylic acids	—C(=O)—OH	-oic acid	carboxy-
carboxylic acid anhydrides	—C(=O)—O—C(=O)—	-oic anhydride	
carboxylic acid esters	—C(=O)—O—	-oate	alkoxycarbonyl- or carboalkoxy-
acyl halides	—COX	-oyl halide	haloalkanoyl-
amides	—CONH$_2$	-amide	carbamoyl-
nitriles	—C≡N	-nitrile or -onitrile	cyano-
aldehydes	—C(=O)—H	-al	alkanoyl-
ketones	—C(=O)—	-one	oxo-
alcohols	—C—OH	-ol	hydroxy-
mercaptans	—C—SH	-thiol	mercapto-
amines	—N<	-amine	amino- or aza-
ethers	—O—	-ether	oxa- or alkoxy-
sulfides	—S—	-sulfide	alkylthio- or thia-
alkenes	C=C	-ene	alkenyl-
alkynes	C≡C	-yne	alkynyl-
halides	—X		halo-
nitro	—NO$_2$		nitro-
alkanes	—C—C—	-ane	alkyl-

Abbreviations are often used to simplify the names of common structural groups or to represent general types of structures. When writing chemical equations, such abbreviations should never be used to represent the atoms at which the reaction takes place. Table 6.3 below summarizes some common structural abbreviations.

Table 6.3　Some common abbreviation

Abbreviation	Groups	Abbreviation	Groups
R-	any alkyl group	Bu-	butyl group ($CH_3CH_2CH_2CH_2-$)
Ar-	any aryl (aromatic) group	iPr-	isopropyl group [$(CH_3)_2CH-$]
X-	any halogen atom	pri-	primary
Ph- or Φ-	benzene substituent (C_6H_5-)	sec-	secondary
Me-	methyl group (CH_3-)	tert-	tertiary
Et-	ethyl group (CH_3CH_2-)	o-	ortho
Pr-	propyl group ($CH_3CH_2CH_2-$)	m-	meta
Ac-	acetyl group (CH_3CO-)	p-	para
py	pyridine	en	ethylenediamine

Unit 7

Introduction to Analytical Chemistry

7.1 Overview

Analytical chemistry is the discipline of chemistry that studies the chemical composition of materials and develops the tools used to study chemical composition. It includes both wet laboratory chemistry and the use of instrumentation. Analytical chemistry is important in science, engineering, medicine, and industry.

Is there iron in moon dust? How much aspirin is in a headache tablet? What trace metals are there in a tin of tuna fish? What is the purity and chemical structure of a newly synthesized compound? These and many other questions about the composition and structure of matter fall within the realm of analytical chemistry. The answers can be provided by simple chemical tests or by the use of expensive and complex instrumentation. The techniques and methods used and the problems encountered are so diverse that they cut across the traditional boundaries of inorganic, organic, and physical chemistry and include aspects of fields such as biochemistry, physics, engineering, and economics[1]. Analytical chemistry is therefore a subject that is broad in scope and requires a specialized and disciplined approach. An inquisitive and critical mind, a keen sense of observation and the ability to pay attention to detail are desirable qualities in anyone who wishes to master the subject. However, it is increasingly recognized that the role of the analytical chemist is not to be tied to a bench with a burette and balance, but to become involved in the broader aspects of the analytical problems encountered. Discussions with scientific and commercial colleagues, customers, and other interested parties, as well as site visits, can greatly assist in the selection of methods and the interpretation of analytical data, thereby minimizing time, effort, and expense.

Analytical chemistry uses standards and error analysis. If chemistry is aptly described as "what chemists do", then analytical chemistry is what analytical chemists do. However, defining analytical chemistry is no easier than describing chemistry as a whole. Biologists, geophysicists, engineers — they all practice chemistry to some extent, and almost all chemists practice analytical chemistry; for chemistry is, among other things, the study of the composition and behavior of the natural world. Anyone who wants to know more about the composition of substances must use analytical methods to determine the types and amounts of compounds, elements, atoms, and subatomic particles present in a given sample, and to study the detailed compositions of the various species[2]. Analytical methods must be used to study the behavior of materials before, during, and after certain reactions or "changes".

A chemical analysis provides information about the composition of a sample of matter. The results of some analyses are qualitative, providing useful information about the molecular or atomic species, structural features, or functional groups in the sample. Other analyses are quantitative; the results take the form of numerical data in units such as percent, parts per million, or milligrams per liter. In both types of analysis, the desired information is obtained by measuring a physical property that is characteristic of the constituent(s) of interest.

It is convenient to describe properties useful for determining chemical composition as analytical signals; examples of such signals include emitted or absorbed light, conductance, mass, volume, and refractive index. None of these signals is unique to a particular species. For example, all metallic elements in a sample typically emit ultraviolet and visible radiation when heated to a sufficiently high temperature in an electric arc; all charged species conduct electricity; and all components in a mixture contribute to its refractive index, mass, and volume[3]. Therefore, all analyses require separation. In some cases, the separation step involves physically isolating the individual chemical components in the sample prior to signal generation; in other cases, a signal is generated or observed for the entire sample and the desired signal is isolated from the others; some signals are susceptible to the latter treatment while others are not. For example, when a sample is heated in an electric arc, the wavelength distribution of radiation is different for each metallic species.

7.2 The Scope of Analytical Chemistry

The boundaries of analytical chemistry are among the broadest of any technological discipline. An analyst must be able to design, perform and interpret measurements in the context of the fundamental technological problem with which he or she is faced. The selection and application of appropriate chemical procedures requires a broad knowledge of chemistry, while familiarity with and ability to operate a wide variety of instrumentation is essential. Finally, the analyst must have a sound knowledge of the statistical treatment of experimental data in order to evaluate the significance and reliability of the results obtained.

When an investigation is limited to the identification of one or more constituents of a sample, it is called qualitative analysis, whereas an investigation to determine how much of a particular species is present is called quantitative analysis. Sometimes information on the spatial arrangement of atoms in a molecule or crystalline compound is required, or confirmation of the presence or position of certain organic functional groups is sought. Such studies are called structural analysis and may be considered more detailed forms of analysis. All species that are the subject of either qualitative or quantitative analysis are called analytes.

There are significant similarities between the techniques and methods utilized in qualitative and quantitative analysis. In both instances, a sample undergoes physical and chemical "conditioning" before measuring a relevant property associated with the analyte. The main difference lies in the degree of control over the relationship between a measurement and the amount of analyte present. For a thorough qualitative analysis, it is recommended to utilize a test with a clearly defined sensitivity limit to accurately evaluate negative and positive results. In a quantitative analysis, however, the relationship between measurement and analyte must follow a strict and measurable proportionality; only then can the amount of analyte in the sample be derived from the measurement. To maintain this proportionality, it is generally essential that all reactions used to prepare a sample for measurement are controlled and reproducible, and that the conditions of measurement remain constant for all similar measurements. Careful calibration of the methods used in quantitative analysis is also essential. These aspects of chemical analysis are a major concern of the analyst.

7.3 The Functions and Patterns of Analytical Chemistry

7.3.1 Functions of Analytical Chemistry

Chemical analysis is an indispensable servant of modern technology, while at the same time it is partly dependent on modern technology for its operation. Indeed, the two have developed hand in hand. Since the early days of quantitative chemistry in the second half of the 18th century, chemical analysis has provided an important foundation for the development of chemistry. For example, the studies of combustion and the atomic theory proposed by Dalton were based on quantitative analytical evidence.

The transistor is a more recent example of an invention that would have been almost impossible to develop without sensitive and accurate chemical analysis. This example is particularly interesting because it illustrates the synergistic development so often observed in different fields. After supporting the development of the transistor, analytical instrumentation is now making extremely wide use of it. It is impossible to overestimate the importance of analysis in modern technology. Some of the most important applications are listed below.

(1) Basic research. The first steps in unraveling the details of an unknown system often involve the identification of its components by qualitative chemical analysis. Subsequent investigations usually require structural information and quantitative measurements. This pattern is found in a wide range of areas such as the creation of novel medications, the investigation of meteoroids, and the analysis of heavy ion bombardment outcomes in nuclear physics.

(2) Product development. The design and development of a new product often depend on establishing a relationship between its chemical composition and its physical properties or performance. Typical examples include the development of alloys and polymer composites.

(3) Product quality control. Most manufacturing industries require consistent product quality. To ensure that this requirement is met, both raw materials and finished products are subjected to extensive chemical analysis. On the one hand, the necessary ingredients must be kept at optimal levels, and on the other hand, contaminants, such as toxins in food, must be kept below the maximum level allowed by law.

(4) Pollutant monitoring and control. Residual heavy metals and organochlorine pesticides are two well-known pollution problems. Sensitive and accurate analysis is required to assess the distribution and level of a pollutant in the environment, and routine chemical analysis is important in the control of industrial effluents.

(5) Assay. In commercial dealings with raw materials such as ores, the value of the ore is determined by its metal content. Often large quantities of material are involved, so that in the aggregate small differences in concentration can be of considerable commercial significance. Accurate and reliable chemical analysis is therefore essential.

(6) Medical and clinical studies. The levels of various elements and compounds in body fluids are important indicators of physiological disorders. High levels of glucose in urine, which indicate the presence of diabetes, and increased levels of lead in the blood, which suggest lead poisoning, are both well-established examples.

7.3.2 General Patterns of Analytical Chemistry

The solutions to all analytical problems, both qualitative and quantitative, follow the same basic pattern. This can be described under seven general headings.

(1) Method selection. The selection of the analytical method is an essential step in solving an analytical problem. A decision should only be made once the overarching issue has been clearly defined, and ideally, the client and analyst should collaborate in the decision-making process. The method selected will inevitably involve a compromise between the desired sensitivity, precision and accuracy of the results and the cost involved. For example, X-ray fluorescence spectrometry may rapidly provide quantitative results for trace element problems, but they may lack precision. Atomic absorption spectrophotometry, on the other hand, will provide more accurate data, but at the expense of more time-consuming chemical manipulations.

(2) Sampling. Proper sampling is essential for reliable analysis. The analyst, with the assistance of their technical colleagues, must determine the appropriate time, location, and method of obtaining a sample that accurately represents the parameter being measured.

(3) Preliminary sample treatment. For quantitative analysis, the amount of sample taken is usually measured by mass or volume. If a homogeneous is available, it can be divided without further treatment. However, solids like ores require crushing and mixing. Additional sample preparation is often necessary for analysis, including drying, ignition, and dissolution.

(4) Separations. Analytical measurements are often subject to interference from other sample components. Modern methods increasingly employ instrumental techniques to distinguish between analyte and interfering signals; however, such discrimination is not always feasible. In some cases, a selective chemical reaction can be used to mask the interfering substance. Nevertheless, separation of the analyte from the interfering component is necessary when this approach fails. If quantitative measurements are to be performed, the separations must also be quantifiable or provide a well-known recovery of the analyte.

(5) Final measurement. This final step is typically the quickest and easiest of the seven, but its accuracy relies directly on the previous steps. The basic requirement is a known proportionality between the magnitude of the measurement and the amount of analyte present. A wide range of parameters can be measured.

(6) Method validation. There is no point in performing the analysis unless it is known that the results obtained are meaningful. This can only be ensured by properly validating the method prior to its use and regularly monitoring its performance. Adhering to validated standards is the most satisfactory approach. Validated standards have undergone extensive analysis using various methods to obtain an accepted value for the appropriate analyte. To ensure accuracy, it is recommended to select a standard with a comparable matrix to that of the sample. To maintain accurate analysis, standards must be periodically reanalyzed.

(7) Evaluation of results. The results of an analysis shall be evaluated by appropriate statistical methods and their significance shall be considered in the light of the original problem.

7.4 Analytical Methods

It is common to find analytical methods classified as classical or instrumental, the former including "wet chemical" methods such as gravimetry and titrimetry. This classification is largely artificial and historically derived, as there is no fundamental difference between the methods used in both groups. All analytical methods involve the correlating a physical measurement with the concentration of the analyte. Few methods, in fact, are purely instrumental, with the most requiring chemical manipulation prior to the instrumental measurement.

7.4.1 Types of Analytical Methods

Table 7.1 presents the most frequently used signals for analytical purposes. Note that the first six signals involve radiation emission or interaction with matter, while the next three signals relate to electricity. The last group of signals contains five miscellaneous signals. The table also lists the names of the analytical methods that are based on each signal.

Table 7.1 Some analytical signals

Signal	Analytical methods based on measurement of signal
emission of radiation	emission spectroscopy (X-ray, UV, visible, electron), flame photometry, fluorescence (X-ray, UV, visible), radiochemical methods
absorption of radiation	spectrophotometry (X-ray, UV, visible, IR), colorimetry, atomic absorption, nuclear magnetic resonance, and electro spin resonance spectroscopy
scattering of radiation	turbidimetry, nephelometry, Raman spectroscopy
refraction of radiation	refractometry, interferometry
diffraction of radiation	X-Ray and electron diffraction methods
rotation of radiation	polarimetry, optical rotatory dispersion, circular dichroism
electrical potential	potentiometry, chronopotentiometry
electrical current	polarography, amperometry, coulometry
electrical resistance	conductimetry
mass-to-charge ratio	mass spectrometry
rate of reaction	kinetic methods
thermal properties	thermal conductivity and enthalpy methods
mass	gravimetric analysis
volume	volumetric analysis

It is interesting to note that until perhaps 1920, most analysis was based on the last two signals listed in Table 7.1, namely mass and volume. As a result, gravimetric and volumetric methods are known as classical analysis (chemical analysis), as opposed to the other methods known as instrumental methods.

Beyond the chronology of their development, there are few features that clearly distinguish instrumental from classical methods. Some instrumental techniques are more sensitive than classical ones, but many are not. For certain combinations of elements or compounds, an instrumental method may be more specific; for others, a gravimetric or volumetric approach may be less subject to interference. Generalizations based on accuracy, convenience, or time are equally difficult to make. Nor is it necessarily true that instrumental methods require more sophisticated or costly equipment; in fact, the use of a modern automatic balance in a gravimetric analysis requires more complex and sophisticated instrumentation than is required for several of the methods listed in Table 7.1.

In addition to the methods listed in the second column of Table 7.1, there is another group of analytical methods for the separation and resolution of closely related compounds. Common separation methods include chromatography, distillation, extraction, ion exchange, fractional crystallization, and selective precipitation. One of the signals listed in Table 7.1 is typically used to complete the analysis following the separation step. For example, thermal conductivity, volume, refractive index, and electrical conductivity have all been used in conjunction with various chromatographic methods.

This text covers most of the instrumental methods listed in Table 7.1, as well as many of the most commonly used separation techniques. Little space is devoted to the classical methods, since their use is usually covered in elementary analytical courses.

Table 7.1 suggests that the chemist faced with an analytical problem may have a bewildering array of methods from which to choose, and the amount of time spent on the analytical work and the quality of the results are critically dependent on this choice. In deciding which method to choose, the chemist must consider the complexity of the materials to be analyzed, the concentration of the species of interest, the number of samples to be analyzed, and the accuracy required. His choice will then depend on a knowledge of the basic principles underlying the various methods available to him, and thus of their strengths and limitations.

7.4.2 Trends in Analytical Methods and Procedures

The techniques and methods of analytical chemistry are constantly evolving and changing. Better instrument design and a better understanding of the mechanics of analytical processes allow continuous improvements in sensitivity, precision, and accuracy. These same changes contribute to more economical analysis, often leading to the elimination of time-consuming separation steps. The ultimate development in this direction is a non-destructive method that not only saves time, but also leaves the sample unchanged for further examination or processing.

Automation of analysis, sometimes using laboratory robots, has become increasingly important. For example, it allows a number of bench analyses to be performed more quickly, efficiently and accurately, while in other cases it allows continuous monitoring of an analyte in a production process. Two of the most important developments in recent years have been the incorporation of microprocessor control into analytical instruments and their interfacing with

Part B Introduction to Fundamentals of Chemistry and Chemical Engineering   113

micro- and minicomputers. The microprocessor has improved instrument control, and performance and, through the ability to monitor the condition of component parts, has simplified routine maintenance. Operation by relatively inexperienced personnel can be facilitated by simple interactive keypad dialogs, including storage and recall of standard methods, report generation, and diagnostic testing of the system. Microcomputers with sophisticated data handling and graphics software packages have also had a significant impact on the collection, storage, processing, enhancement, and interpretation of analytical data. Laboratory Information and Management Systems (LIMS) which automatically record large numbers of samples, chemometrics which involves computerized and often sophisticated statistical analysis of data, and expert systems which provide interactive computerized guidance and evaluation in solving analytical problems, have all become important in optimizing chemical analysis and maximizing the information it provides[4].

Analytical problems continue to emerge in new forms. The need for long-range analysis with instrument packages is constantly increasing. Space probes, well logging, and deep sea studies are examples of these requirements. In other areas, such as environmental and clinical studies, it is increasingly recognized that the exact chemical form of an element in a sample is more important than the mere level of its presence. Two well-known examples are the much greater toxicity of organolead and organomercury compounds compared to their inorganic counterparts. Identification and determination of the element in a specific chemical form presents the analyst with some of the more difficult problems.

7.5 Some Modern Methods of Analytical Chemistry

7.5.1 High Performance Liquid Chromatography

Liquid chromatography was discovered in 1903 by M. S. Tswett, who used a chalk column to separate pigments from green leaves. It was not until the 1960's that more emphasis was placed on the development of liquid chromatography. High Performance Liquid Chromatography (HPLC) is a type of chromatography that is the most widely used analytical technique. Chromatographic processes can be defined as separation techniques involving mass transfer between stationary and mobile phases. HPLC uses a liquid mobile phase to separate the components of a mixture. These components (or analytes) are first dissolved in a solvent and then forced to flow through a chromatographic column under high pressure. In the column, the mixture is separated into its components. The amount of resolution is important and depends on the amount of interaction between the dissolved components and the stationary phase. The stationary phase is defined as the immobile packing material in the column. The interaction of the solute with the mobile and stationary phases can be manipulated by different choices of solvents and stationary phases, giving HPLC a high degree of versatility not found in other chromatographic systems and the ability to easily separate a wide variety of chemical mixtures[5]. The basic components of an HPLC system include a solvent reservoir, pump, injector, analytical column, detector, recorder, and waste reservoir, as shown in Figure 7.1.

An HPLC system begins with the solvent reservoir, which contains the solvent used to move the sample through the system. The solvent should be filtered with an inlet solvent filter to remove

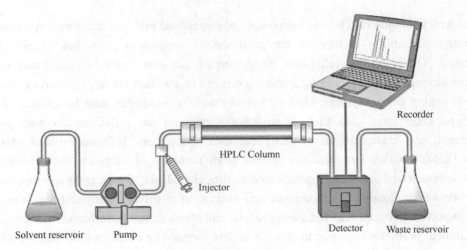

Figure 7.1 Basic HPLC system

any particles that could potentially damage the sensitive components of the system. The solvent is carried through the system by the pump. This often includes internal pump seals that slowly degrade over time. As these seals break down and release particles into the flow path, an in-line solvent filter prevents any damage to the post-pump components. The next component in the system is the sample injector, also known as the injection valve. This valve, equipped with a sample loop of the appropriate size for the analysis being performed, allows for reproducible introduction of sample into the flow path. Because the sample often contains particulate matter, it is important to use either a sample filter or a precolumn filter to prevent damage to the valve and column. Following the injector, an analytical column provides primary sample separation. This is based on the differential attraction of sample components to the solvent and packing material within the column. Following the analytical column, the separated components pass through a detector flow cell before entering the waste reservoir. The presence of sample components in the flow cell triggers an electrical response from the detector, which is digitized and sent to a recorder. The recorder assists in the analysis and interpretation of the data. As a final system enhancement, a back pressure regulator is often installed immediately after the detector. This device prevents the formation of solvent bubbles until the solvent has completely passed through the detector. This is important because bubbles in a flow cell can interfere with the detection of sample components. Alternatively, an inert gas sparging system can be installed to force dissolved gases out of the solvent stored in the solvent reservoir.

Consider the separation of a two-component mixture dissolved in the eluent. Assume that component A has the same interaction with the adsorbent surface as an eluent and component B has a strong excess interaction. When injected into the column, these components are driven through by the eluent flow. The molecules of component A will interact with and retard the adsorbent surface and retard on it in the same way as an eluent molecule. The average result is that component A will move through the column at the same rate as an eluent. Molecules of component B that are adsorbed to the surface (due to their strong excess interactions) will remain on the surface for much longer. As a result, it will move through the column more slowly than the eluent flow. A general shape of the chromatogram for the mixture is shown in Figure 7.2.

Usually, a relatively narrow band is injected (5 - 20 μL injection volume). During the run, the original chromatographic band is broadened due to uneven flow around and inside the porous particles, slow adsorption kinetics, longitudinal diffusion, and other factors. Together, these processes produce what is known as band broadening of the chromatographic zone. In general, the longer the component is retained on the column, the broader its zone (peak on the chromatogram). Separation performance depends on both component retention and band broadening.

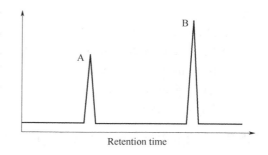

Figure 7.2 General retention time peaks of the chromatogram for the mixture

Band broadening is generally a kinetic parameter that depends on adsorbent particle size, porosity, pore size, column size, shape, and packing performance. Retention, on the other hand, does not depend on the above parameters, but reflects molecular surface interactions and depends on the total adsorbent surface area.

HPLC is the most widely used analytical separation technique today. The method is popular because it is non-destructive and can be applied to thermally labile compounds (unlike GC); it is also a very sensitive technique because it incorporates a wide range of detection methods. By using post-column derivatization methods to improve selectivity and detection limits, HPLC can be easily extended to trace levels of compounds that normally do not provide adequate detector response. The wide applicability of HPLC as a separation method makes it a valuable separation tool in many scientific fields.

Liquid chromatography is useful for a variety of applications in industry and academia. Its use can be divided into two classifications, analytical LC and preparative LC. In analytical LC, the goal is to identify and quantify specific components within a sample, typically in the picogram to milligram range. In preparative LC, the goal is to isolate or collect the separated components of the sample in the mg to kg range.

In industry and science, analytical LC is used for:
✧ Basic research
✧ Quality assurance
✧ Methods development

Liquid chromatography is widely used in different types of industry:
✧ Life science, including proteins, nucleic acids, carbohydrates, lipids, metabolites
✧ Pharmaceuticals
✧ Biotechnology
✧ Industrial chemicals, including fine chemicals, polymers, synthetic mixtures
✧ Food and agricultural processing, including plant products, agrochemicals

The preparative capabilities of LC are applied at different levels or scales of isolation:
✧ Small-scale or semi-preparative (mg to g)
✧ Pilot-scale (g to kg)
✧ Production or process scale (kg to ton)

In preparative LC, a partial list of specific needs or reasons to isolate or purify samples is as

follows:
- ◇ Drug efficiency studies
- ◇ Full-scale production of drugs (e.g., interferon, insulin)
- ◇ Spectroscopy/structure elucidation
- ◇ Biological screening
- ◇ Physical testing

7.5.2 High Performance Capillary Electrophoresis

Electrophoresis refers to the migration of charged electrical species when dissolved or suspended in an electrolyte through which an electrical current is passed. Cations migrate to the negatively charged electrode (cathode) and anions are attracted to the positively charged electrode (anode). Neutral solutes are not attracted to either electrode. Traditionally, electrophoresis has been performed on layers of gel or paper. Traditional electrophoresis equipment offered a low level of automation and long analysis times. Detection of the separated bands was done by visualization after separation. Analysis times were long because only relatively low voltages could be applied before excessive heat generation caused loss of separation.

The advantages of capillary electrophoresis (CE) were highlighted in the early 1980s by the work of Jorgenson and Lukacs, who popularized the use of CE. Capillary electrophoretic separations were shown to offer the possibility of automated analysis equipment, fast analysis times, and on-line detection of separated peaks. The heat generated inside the capillary was effectively dissipated through the capillary walls, allowing high voltages to be used to achieve rapid separations. The capillary was inserted through the optical center of a detector, allowing on-capillary detection.

Capillary electrophoresis has evolved into a collection of separation techniques in which high voltages are applied across buffer-filled capillaries to achieve separations. Variations include separations based on size and charge differences between analytes (called capillary zone electrophoresis, CZE, or free solution CE, FSCE), separation of neutral compounds using surfactant micelles (micellar electrokinetic capillary chromatography, MECC, or sometimes referred to as MEKC), separation of solutes through a gel network (capillary gel electrophoresis, CGE), and separation of zwitterionic solutes within a pH gradient (capillary isoelectric focusing, CIEF). Capillary electrochromatography (CEC) is a related electrokinetic separation technique in which voltages are applied across capillaries filled with silica gel stationary phases. Separation selectivity in CEC is a combination of both electrophoretic and chromatographic processes. Many of the CE separation techniques rely on the presence of an electrically induced flow of solution (electroosmotic flow, EOF) within the capillary to pump solutes toward the detector.

FSCE and MECC are the most commonly used separation techniques in pharmaceutical analysis. CGE and CIEF are important for the separation of biomolecules such as DNA and proteins, respectively, and are becoming increasingly important with the development of biotechnology-derived drugs increases. In general, CE is generally performed using aqueous-based electrolytes, but there is an increasing use of non-aqueous solvents in CE.

The operation of a CE system involves the application of a high voltage (typically 10 - 30 kV) through a narrow bore (25 - 100 μm) capillary. The capillary is filled with electrolyte solution which conducts the current through the interior of the capillary. The ends of the capillary are immersed in electrolyte-filled reservoirs. Electrodes made of an inert material such as platinum are also inserted into the electrolyte reservoirs to complete the electrical circuit. A small volume of sample is injected into one end of the capillary. The capillary passes through a detector, usually a UV absorbance detector, at the opposite end of the capillary. The application of a voltage causes the sample ions to move toward their corresponding electrode, usually through the detector. A plot of the detector response versus time is produced, called an electropherogram. A flow of electrolyte, known as electroosmotic flow (EOF), causes the solution to flow along the capillary, usually toward the detector. This flow can significantly reduce analysis times or force an ion to overcome its tendency to migrate toward the electrode to which it is attracted by the sign of its charge.

Detailed treatments of the background theory and non-pharmaceutical applications can be found in a number of reference books.

Commercially available CE instruments, as shown in Figure 7.3, are PC-controlled and consist of a buffer-filled capillary that passes through the optical center of a detector, a means of introducing the sample into the capillary, a high voltage power supply and an autosampler.

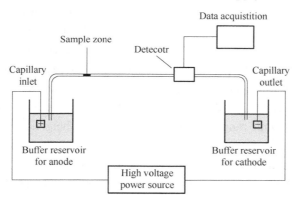

Figure 7.3 Typical CE separation system

Typical voltages used are in the range of 5 - 30 kV, resulting in currents in the range of 10 - 100 µA. Higher currents can cause problems with heating in the capillary, which can broaden peaks and lead to loss of resolution.

Capillaries

The capillaries used are typically fused silica capillaries coated with an outer polyimide protective coating to provide increased mechanical strength, as bare fused silica is extremely fragile. A small portion of this coating is removed to form a window for detection purposes. The window is aligned with the optical center of the detector. Capillaries are typically 25 - 100 cm long, with 50 and 75 microns being the most commonly used inner diameters. Capillary volumes are on the order of 1 µL, e.g. the approximate volume of a 50 µm wide, 50 cm long capillary is 1 µL. The volume can be calculated as volume = $\pi L d^2/4$. In standard commercial CE instruments, the capillary is often held in a housing such as a cartridge to facilitate easy insertion of the capillary into the instrument and to protect the sensitive detection window area. The inner surface of the capillary can be chemically modified by covalently bonding (coating) various substances onto the capillary wall. These coatings are used for a variety of purposes, such as reducing sample adsorption or changing the ionic charge on the capillary wall.

Temperature control

It is important to control the temperature of the environment around the capillary to ensure consistent separations. To achieve this, capillaries are often inserted into cartridges that are placed inside the CE instrument. Temperature controlled air or liquid coolant is then forced through the cartridge to regulate the temperature.

Sample injection

The sample solution is injected into the end of the capillary furthest from the detector. Typical injection volumes are 10 - 100 nL. The most commonly used injection mode is to dip the capillary into the sample solution vial. The vial is then pressurized to force a volume of solution into the capillary. An alternative, less popular sample injection method is to dip the capillary and electrode into the sample solution vial and apply a voltage. If the sample is ionized and the proper voltage polarity is used, sample ions will migrate into the capillary. This type of injection is known as electrokinetic sampling.

Detectors

The most commonly used detector is a UV absorption detector, which is standard on commercial CE instruments. UV diode detectors are also available on most instruments. Alternative commercially available detector modes include fluorescence, laser-induced fluorescence, conductivity, and indirect detection. The combination of CE and mass spectrometry is often used to provide structural information on the resolved peaks. Detectors can be interfaced with data acquisition systems to calculate results. Integrated peak areas are routinely used for quantitation as they provide increased dynamic ranges compared to using peak heights.

Power supply

Separations are typically performed using voltages in the range of 5 - 30 kV. Electrolyte ionic strengths are generally chosen during method development so that the application of these voltages produces currents of 10 - 100 μA. Operation with currents above this level can result in unstable, non-reproducible operating conditions. On many instruments it is possible to operate by applying constant voltage (most common), constant current, or constant power across the capillary. However, constant voltage is the most commonly used mode of operation.

7.5.3 Infrared Spectrophotometry

Much of what has been said about spectrophotometry in the ultraviolet and visible regions of the spectrum can be directly applied to the infrared, since the basic principles are the same. However, there are several significant differences at the practical level. One obvious difference is in the nature of the process being observed. In the infrared, vibrational rather than electronic transitions are observed. The lattice vibrations of ionic crystals are broad and not important for analytical applications. Therefore, we limit the discussion to compounds with covalent bonds, especially the vibrations of organic substances. Not all covalently bonded molecules give infrared absorption spectra; a change in the dipole moment is also required. For example, molecules such as H_2, N_2, or O_2 do not undergo a change in dipole moment and do not give an infrared spectrum.

Electronic absorption spectra tend to contain few transitions; often only a single one is important. Especially in solutions, electronic transitions are broadened by numerous vibrational, rotational, and solvent interactions, resulting in broad spectral bands. In the infrared, where vibrational spectra are observed, the transitions are usually not broadened, but remain relatively narrow. Because an infrared spectrum can contain a large number of narrow absorption bands, it is likely to be highly specific. For example, a methyl functional group has CH stretching, CH_2 bending, hydrogen-hydrogen scissoring, and twisting vibrational modes, in addition to a carbon-carbon stretching vibration when the carbon in the methyl group is bonded to another carbon. Each of these vibrational transitions is characteristic of a methyl group.

The frequency of a vibrational absorption band increases with increasing bond energy and decreasing atomic mass. Vibrational energy absorption does not involve a shift in the center of mass of a molecule, which would only involve a trans-dimensional movement of the entire molecule. Each vibrational absorption is accompanied by several rotational levels of only slightly different energies; thus, an absorption band is observed rather than a narrow absorption line. In infrared measurements, a non-polar solvent is usually chosen to reduce interference from solvent absorption. There are several vibrational modes that can occur in a molecule of carbon dioxide — symmetric stretching, scissoring, and asymmetric stretching. For more complex molecules, some of the other vibrational modes of energy transfer are described as twisting, wiggling, rocking, and so on.

Infrared spectra are usually plotted as transmittance versus wavelength or wave number.

7.5.4 Nuclear Magnetic Resonance

Deep inside the submicroscopic atom is the even tinier nucleus. Many atomic nuclei behave like spinning micromagnets; when placed in an external magnetic field, these micromagnets attempt to align themselves either parallel or antiparallel to the field. This behavior is analogous to that of a gyroscope in the Earth's gravitational field. If a radio frequency field of just the right energy is applied to these nuclei, the energy can be absorbed by the nuclear micromagnets and cause them to move from one orientation to the other. Nuclear magnetic resonance(NMR) was first observed in 1946. Although all nuclei have both charge and mass, only about half of the isotopes of the elements have a spinning nucleus that generates a small magnetic field.

The most commonly used isotopes are those of hydrogen, fluorine-19, nitrogen-15, phosphorus-31, and carbon-13. By far the most information has been obtained with hydrogen nuclei (protons) due to the sensitivity of the proton NMR signal. Since hydrogen is present in virtually all organic compounds, it is a useful general probe. Carbon-13 measurements are also valuable for understanding the structural details of organic molecules.

The energy required to produce an NMR signal corresponds to a wavelength of the order of 1 cm and a frequency of 10^8 Hz (100 MHz), about 100,000 times lower than the vibrational energies involved in the infrared region. The detector must be sensitive to these low frequencies, which are in the radio frequency range.

From the magnetic moment of the spinning nucleus, it can be calculated that a proton in a magnetic field of 14092 Gauss magnetic field will absorb or emit energy at a frequency of 60 MHz; this is called the Larmor frequency for the particular field and nucleus. All hydrogen nuclei in a 14092 Gauss magnetic field absorb energy at approximately this frequency. The slight variations in absorption frequency that occur are due to small differences in the actual magnetic fields around the different nuclei due to small secondary fields generated by the electrons surrounding them. These small locally induced fields partially overcome the effect of the much larger external field imposed on the nuclei. Thus, they respond to magnetic fields somewhat less than would be expected from the magnitude of the applied field. The induced field depends on the electron density around the nuclei; the higher the density, the stronger the field. Because its magnitude depends on the chemical environment, the change in absorption frequency due to the induced field is called a chemical shift. Chemical shifts are often given as relative values for convenience.

The phenomenon of chemical shifts in nuclear resonance was discovered in 1949, and by 1956 chemists had become the main beneficiaries of NMR. It is so important to chemistry because the chemical shift of the nucleus of an atom, such as a hydrogen atom, depends on the particular molecule in which the hydrogen atom is present. NMR has revolutionized the identification and characterization of molecules.

Chemical shifts are minute; at 10^8 Hz, proton chemical shifts can cover a range of 1000 Hz. It is often desirable to measure chemical shifts on the order of 0.1 Hz, which at 10^8 Hz means a measurement accurate to 1 part in 10^9. This places high demands on the instrumentation.

An NMR spectrum provides information about (1) chemical shifts, which reflect the chemical environment, (2) spin-spin splitting, and (3) peak areas. Spin-spin splitting results from the effects of nearby protons on the energy levels of the protons being observed. For example, in a molecule containing the ethyl group, $CH_2CH_2^-$, the methyl protons appear as three closely spaced absorption peaks rather than a single peak due to splitting by the CH_2 protons.

Peak areas, or absorptions, indicate the number of protons of a particular type or provide the basis for quantitative measurements in mixtures. Qualitative NMR analysis uses all three types of information and is enormously powerful for identifying the structures of organic compounds.

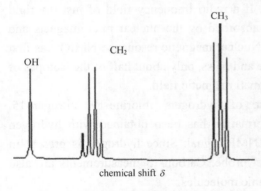

Figure 7.4 NMR spectrum of ethanol

A simple example of a proton NMR spectrum is that of ethanol (CH_3CH_2OH). Ethanol contains hydrogen atoms in three chemically distinct environments, each of which absorbs radio frequency radiation at slightly different frequencies. The amount of absorption, when plotted as a function of frequency, provides information about the immediate environment of that particular hydrogen nucleus. Figure 7.4 shows the proton NMR spectrum of ethanol.

The basic instrument components for NMR measurements are (1) a magnet to provide a uniform magnetic field, (2) a radio frequency transmitter and receiver, (3) a sample holder, and (4) the necessary associated electronics for data acquisition and processing. Modern instruments require sophisticated electronics to control the radio frequency transmission, detect the absorbed frequencies, amplify the signals, and present the information to the operator. Data acquisition systems may include a minicomputer to collect and process the radio frequency signals to extract maximum chemical information.

To measure frequency differences of 1 part in 10^9, the magnetic field must be both constant and homogeneous. Better electronic stability, and therefore better accuracy, is achieved by keeping the frequency constant, rather than varying it, and scanning by slightly varying the total magnetic field using sweep coils. When the magnetic field is varied, a spectrum is obtained. Since the total magnetic field is not known exactly, the chemical shifts are measured relative to a reference. An example of a reference is $(CH_3)_4Si$. The electron density around the hydrogen nuclei of $(CH_3)_4Si$ is greater than that around the hydrogen nuclei in most organic compounds, thus providing an isolated line.

The magnet is the heart of the instrument. Homogeneity of the magnetic field is essential; it is improved by spinning the sample to average out inhomogeneities. To observe small differences in radio frequency absorption, the magnetic field should be both large and uniform. With conventional electron magnets, the largest field that can be developed is about 23000 Gauss, an immense value. The generation of even stronger magnetic fields has recently become practical with the development of superconducting magnets. Special alloys of neodymium and titanium at liquid helium temperatures lose all resistance to the flow of electricity even in a large magnetic field. A large current can then flow without generating heat. Thus, larger currents and magnetic fields are possible with superconductivity. Magnetic fields are now achievable that are about four times the maximum of conventional electromagnets (and about a million times stronger than the Earth's magnetic field). These larger magnetic fields have increased the spread of frequencies, thereby improving resolution, while providing higher sensitivity with small amounts of material. Combined with strong fields, the vastly increased sophistication and reliability of today's electronics and computers allow the study of more complex molecules and smaller amounts of material.

Many NMR instruments now use the Fourier Transform (FT) mode of operation, in which the sample is irradiated at a number of frequencies simultaneously at high intensity for a short time, such as 50 μs. The radiation is then turned off, and as the sample returns to equilibrium, a signal is generated as a function of time. The advantage of FT is that a complete spectrum can be obtained in about a second instead of several minutes.

Approximately 1% of naturally occurring carbon is carbon-13. For this reason, and because the sensitivity of the carbon-13 nucleus is less than that of the proton, the overall sensitivity of carbon-13 measurements is about 0.1% of the sensitivity of proton NMR. The development of FT techniques has made it possible to obtain high quality carbon-13 spectra. The information from carbon-13 NMR spectra is complementary to that from proton NMR. For example, decane contains many CH_2 and CH_3 protons, giving a proton spectrum of little value for structural studies. With carbon-13 NMR, each carbon in the decane chain produces distinct, well-separated peaks.

7.5.5 Atomic Absorption

Knowledge of the phenomenon of atomic absorption (AA) has a long history. The Fraunhofer lines in the solar spectrum, caused by the absorption of radiation by cooler atoms in the outer regions, were observed more than a century and a half ago. However, it was not until 1955 that Walsh proposed that atomic absorption could be used for quantitative analysis.

As the name implies, the analytical information is encoded by an absorption process. The flame is placed between a radiation source of optimal wavelength and a detector in a manner analogous to a spectrophotometric cell. The sample solution is first drawn into the burner nebulizer by a stream of air or other oxidant such as nitrous oxide and then mixed with fuel. Most of the sample solution enters the nebulizer chamber as relatively large droplets that settle and then pass down the drain tube. The remaining solution is carried by the air-fuel mixture as a mist to the burner head, where the solvent evaporates and the solute is dissociated into atoms by the heat of the flame. The number of atoms that reach this point in the process is a small fraction of the total. For high sensitivity, as many of these atoms in the flame as possible should absorb radiation from the source. The ideal source for this purpose should be of high intensity at the wavelength required for the element to be determined and of low intensity at all other wavelengths. The closest approximation to this ideal is a lamp whose cathode contains the element to be determined. When heated, atoms of that element will emit energy at the wavelengths most likely to be absorbed by the same element in the flame. This provides both selectivity and sensitivity, since other elements in the sample generally do not absorb radiation close enough to the selected wavelength to interfere with the measurement. Background interference is reduced by inserting a monochromator with a grating (or sometimes a quartz prism) between the sample and the detector.

When a beam of light from a radiation source is directed to pass through a sample in the form of an atomic vapor, some of the radiation is absorbed and some is transmitted. The amount absorbed is proportional to the concentration of atoms of interest in the high-temperature vapor. In all emission techniques, the sample is excited, and the resulting radiation is measured. In atomic absorption, however, the element of interest is not excited in the flame. Instead, it is converted to ground state atoms that can further absorb radiation of the same frequency that would be emitted if the element were excited. Dissociation is typically achieved by burning the sample in a flame or heating it in a small furnace, which excites a small fraction of the atoms. At a temperature of 3000°C, only one out of approximately 10^9 zinc atom is not in the ground state. In flames at high temperatures, up to 2% of the sodium atoms may be excited, but the proportion of ground state atoms remains in a relatively stable state, allowing high precision measurements.

The instrument for atomic absorption analysis typically consists of a source lamp, chopper, flame, monochromator, and detector. The apparatus is arranged so that the detector alternately reads a signal for first the lamp plus flame and then for the flame alone. The signal can be related to the amount of light absorbed by the elements in the flame. In this instrument, the flame gas solution replaces the conventional solution of conventional spectrophotometric analysis. Cutting off the radiation from the hollow cathode lamp is necessary to eliminate the emission from excited atoms in the flame, which would otherwise produce a background that varies with concentration. Either mechanical (rotating shutter) or electrical (pulsed hollow cathode output) chopping can be used.

The source lamp in an atomic absorption instrument must provide radiation of sufficient intensity to excite a significant number of the ground states of the atoms of interest to higher levels. It should also, if possible, avoid excitation of atoms of other elements in the same wavelength region. Both goals are admirably achieved by the use of a hollow cathode lamp. These lamps are low-pressure, low-temperature, cold cathode tubes that produce emission lines with spectral bandwidths on the order of 10^{-5} nm. The spectral bandwidth of an absorption line in the flame is on the order of 10^{-3} nm, which is broad compared to the emission line of the source. The cathode contains the same element as the one to be determined in the sample. When a current flows between the cathode and anode, positive ions of the metal are stripped from the cathode surface. These ions collide with neon or argon atoms in the lamp, leaving the metal atoms in an excited state from which they can emit their characteristic radiation when they return to the ground state. Such lamps can produce an extremely intense source at precisely the wavelength that can be absorbed by the atoms in their ground state. Spectral interference is thus minimal. Nevertheless, a monochromator is included to isolate the particular line whose intensity is reduced, just as a monochromator isolates a wavelength region in conventional spectrophotometry. The monochromator in an AA system rejects other lines and reduces flame background emission.

Atomic absorption spectroscopy is particularly well suited for measuring small amounts of elements, usually metals, in a sample. It is exceptionally simple, sensitive, and selective.

The typical procedure for dissociating an element from its original matrix involves dissolving the sample and then passing the solution through a flame. To improve sensitivity, a small sample portion may be placed in a graphite tube, heated to evaporate any solvents, and then rapidly heated in an electric furnace to vaporize the sample. The vapor is then sent through the optical path of an AA instrument where the absorbance of the flame is measured. Solutions with a concentration greater than 10^{-3} to 10^{-5} mol/L should be diluted to fall within this range. This can be accomplished with just a few milliliters of solvent.

Atomic absorption techniques can determine numerous elements, replacing slower, more arduous methods in the analysis of trace metals within various organic, inorganic, and biological systems. The technique proves exceptionally satisfactory in determining the concentration ranges of magnesium in cast iron from 0.002% to 0.1%, and silver, zinc, copper, and lead in cadmium metal from 0.004% to 0.4%. Beer's Law describes quantitative absorption if the gas flow rate, solute ion viscosity and surface tension, and nebulizer power remain constant and reproducible.

Although atomic absorption has developed rapidly as a convenient analytical method, it must be used with care. Instrument drift due to changes in lamp intensity and monochromator settings over time can affect results. Also, because of the extreme sensitivity of the method, trace contamination from reagents, water, and the environment during sample preparation and analysis can cause problems. Frequent checking of instrument performance with standard solutions and careful preparation and handling of standards and samples are essential.

As with flame emission, non-spectral sources of interference can become important. Evaporation interferences, such as that of phosphate in the determination of calcium, can be reduced by adding another element, such as lanthanum. The lanthanum then binds the phosphate so that the evaporation of calcium is not impeded. Atoms such as sodium or potassium can cause ionization interference, resulting in increased emission from other easily ionized metals. This interference is reduced by adding a non-determining alkali metal to samples and standards.

In summary, atomic absorption and flame emission methods are more selective than titrimetric methods. It is possible to quantitatively analyze for elements in the presence of many others without the need for separation. A disadvantage is that these methods provide only elemental analysis and no information about molecules, since the molecular structure is destroyed in the process.

An important new excitation source for emission spectroscopy is the **inductively coupled plasma (ICP)** discharge. A sample solution is atomized in a stream of inert gas, usually argon, and passed through a quartz tube surrounded by a coil through which a radio frequency is passed. The high frequency dissociates the argon into a stable plasma (a conductive gaseous mixture of ions and electrons). The temperature of the plasma is high (up to 10 000°C). With the high temperature and inert atmosphere of an ICP, chemical interferences are rare. This characteristic, together with the stability of the source, makes it useful for quantitative multi-element analysis. Sensitivity is generally better than with arc or spark sources.

New Words and Expressions

discipline	[ˈdɪsəplɪn]	n. 纪律，学科
composition	[ˌkɒmpəˈzɪʃn]	n. 成分构成，成分
tool	[tu:l]	n. 工具
instrumentation	[ˌɪnstrəmenˈteɪʃn]	n. 使用仪器
iron	[ˈaɪən]	n. 铁
dust	[dʌst]	n. 粉末，灰尘
trace	[treɪs]	v. 发现，追踪，追溯；n. 痕迹，微量
tuna	[ˈtju:nə]	n. 金枪鱼
scope	[skəup]	n. 范围，领域
critical	[ˈkrɪtɪk(ə)l]	adj. 极其重要的，关键的
keen	[ki:n]	adj. 渴望的，热衷的
burette	[bjuˈret]	n. 滴定管，量管
commercial	[kəˈmɜːʃ(ə)l]	adj. 商业的
extent	[ɪkˈstent]	n. 程度，范围
subatomic	[ˌsʌbəˈtɒmɪk]	adj. 亚原子的，原子内的
clue	[klu:]	n. 线索，提示
quantitative	[ˈkwɒntɪtətɪv]	adj. 数量的，定量的（quantifiable adj. 可以定量的，能量化的）
milligram	[ˈmɪlɪɡræm]	n. 毫克
arc	[ɑ:k]	n. 弧，电弧，弧形
constituent	[kənˈstɪtʃuənt]	n. 成分，构成要素
analyte	[ˈænəlaɪt]	n. （被）分析物，分解物，被测物
transistor	[trænˈzɪstə(r)]	n. 晶体管，晶体管收音机
conductance	[kənˈdʌktəns]	n. 电导，导率（conductivity n. 导电性，传导性）
volume	[ˈvɒlju:m]	n. 体积，容积
refractive	[rɪˈfræktɪv]	adj. 折射的
ultraviolet	[ˌʌltrəˈvaɪələt]	n. 紫外光
metallic	[məˈtælɪk]	adj. 含金属的；n. 金属制品
boundary	[ˈbaundri]	n. 分界线，边界
crystalline	[ˈkrɪstəlaɪn]	adj. 透明的
sufficient	[səˈfɪʃ(ə)nt]	adj. 足够的，充足的
calibration	[ˌkælɪˈbreɪʃn]	n. 标定，校准
preoccupation	[priˌɒkjuˈpeɪʃn]	n. 全神贯注，入神
indispensable	[ˌɪndɪˈspensəbl]	adj. 不可或缺的
accurate	[ˈækjərət]	adj. 准确的，精确的

synergistic	[ˌsɪnəˈdʒɪstɪk]	adj. 协同的，协作的
underpin	[ˌʌndəˈpɪn]	v. 支持，加固
meteorite	[ˈmiːtiəraɪt]	n. 陨石，陨星，流星
bombardment	[bɒmˈbɑːdmənt]	n. 轰炸，炮击
manufacturing	[ˌmænjuˈfæktʃərɪŋ]	n. 制造，制造业
ingredient	[ɪnˈɡriːdiənt]	n. 成分，要素，因素
impurity	[ɪmˈpjʊərəti]	n. 杂质
organochlorine	[ɔːˌɡænəʊˈklɔːriːn]	n. 有机氯
pesticide	[ˈpestɪsaɪd]	n. 杀虫剂，农药
ore	[ɔː(r)]	n. 矿石
urine	[ˈjʊərɪn]	n. 尿液，小便
diabetic	[ˌdaɪəˈbetɪk]	adj. 糖尿病的；n. 糖尿病患者
recovery	[rɪˈkʌvəri]	n. 回收，恢复，痊愈
feasible	[ˈfiːzəbl]	adj. 可行的，行得通的
precision	[prɪˈsɪʒ(ə)n]	n. 精确，准确
fluorescence	[fləˈresns]	n. 荧光，荧光性
spectrometry	[spekˈtrɒmɪtri]	n. 光谱测定法
manipulation	[məˌnɪpjuˈleɪʃn]	n. 操纵，控制
conjunction	[kənˈdʒʌŋkʃ(ə)n]	n. 结合，同时发生
ignition	[ɪɡˈnɪʃ(ə)n]	n. 着火，点燃
magnitude	[ˈmæɡnɪtjuːd]	n. 巨大，重要性
validation	[ˌvælɪˈdeɪʃn]	n. 验证，确认
matrix	[ˈmeɪtrɪks]	n. 矩阵，模型
gravimetry	[ɡrəˈvɪmətri]	n. 重量测定，密度测定
titrimetry	[taɪˈtrɪmɪtri]	n. 滴定分析
chronology	[krəˈnɒlədʒi]	n. 年表，年代学
generalization	[ˌdʒen(ə)rəlaɪˈzeɪʃ(ə)n]	n. 概括，泛论
chromatography	[ˌkrəʊməˈtɒɡrəfi]	n. 色谱分析法
distillation	[ˌdɪstɪˈleɪʃn]	n. 精馏，蒸馏，净化
crystallization	[ˌkrɪstəlaɪˈzeɪʃ(ə)n]	n. 结晶化，具体化
thermal	[ˈθɜːm(ə)l]	adj. 热的，热量的
principle	[ˈprɪnsəp(ə)l]	n. 准则，原则
microprocessor	[ˌmaɪkrəʊˈprəʊsesə]	n. 微处理器
maintenance	[ˈmeɪntənəns]	n. 维护，保养，保持
facilitate	[fəˈsɪlɪteɪt]	v. 使更容易，使便利
storage	[ˈstɔːrɪdʒ]	n. 储存，存储
log	[lɒɡ]	n. 正式记录，日志
chemometrics	[ˈkeməʊˈmetrɪks]	n. 化学计量学
optimize	[ˈɒptɪmaɪz]	v. 优化，充分利用
toxicity	[tɒkˈsɪsəti]	n. 毒性
chalk	[tʃɔːk]	n. 粉笔
injector	[ɪnˈdʒektə(r)]	n. 注射器

filter	[ˈfɪltə(r)]	n. 过滤器；v. 过滤，渗入，透过
inlet	[ˈɪnlet]	n. 进口，入口；v. 嵌入，插入
release	[rɪˈliːs]	v. 释放，放走
precolumn	[prɪeːˈkələm]	n. 前置柱
bubble	[ˈbʌb(ə)l]	n. 气泡，泡沫；v. 冒泡，沸腾，忙碌，活跃，兴奋，激动
reservoir	[ˈrezəvwɑː(r)]	n. 水库，蓄水池，储藏
eluent	[ˈeljuənt]	n. 洗脱液，洗提液
porous	[ˈpɔːrəs]	adj. 有气孔的
adsorption	[ədˈzɔːpʃ(ə)n]	n. 吸附
diffusion	[dɪˈfjuːʒn]	n. 扩散，传播
labile	[ˈleɪbɪl]	adj. 易变的，不稳定的
derivatization	[dɪˌrɪvətaɪˈzeɪʃən]	n. 衍生，衍生化
picogram	[ˈpiːkəʊˌɡræm]	n. 微微克，皮克
isolate	[ˈaɪsəleɪt]	v. 孤立，分离，隔离（isolation n. 隔离，绝缘，离析）
migration	[maɪˈɡreɪʃn]	n. 迁徙
voltage	[ˈvəʊltɪdʒ]	n. 电压，伏特
capillary	[kəˈpɪləri]	n. 毛细血管，毛细管
micelle	[mɪˈsel]	n. 胶束
zwitterionic	[ˌzwɪtəraɪˈɒnɪk]	adj. 两性离子的
gradient	[ˈɡreɪdiənt]	n. 斜坡，坡度，梯度
pharmaceutical	[ˌfɑːməˈsuːtɪk(ə)l]	adj. 制药的；n. 药物
electropherogram	[elektrəfərɒɡˈræm]	n. 电泳图（谱）
electrolyte	[ɪˈlektrəlaɪt]	n. 电解液，电解质
electroosmotic	[elektruːzˈmɒtɪk]	n. 电渗透，电渗
silica	[ˈsɪlɪkə]	n. 二氧化硅，硅土
cartridge	[ˈkɑːtrɪdʒ]	n. 弹壳，墨盒
diode	[ˈdaɪəʊd]	n. 二极管
calculate	[ˈkælkjuleɪt]	v. 计算，核算，预测
lattice	[ˈlætɪs]	n. 晶格，格子，格架
vibration	[vaɪˈbreɪʃ(ə)n]	n. 震动，颤动
covalent	[ˌkəʊˈveɪlənt]	adj. 共价的
spectrum	[ˈspektrəm]	n. 范围，幅度，光谱
scissor	[ˈsɪzə]	v. 用剪刀剪；adj. 剪刀的，剪刀似的，n. 剪刀
dimensional	[daɪˈmenʃn(ə)l]	adj. 维度的
stretching	[ˈstretʃɪŋ]	n. 伸长，展宽；v. 伸展，拉紧
submicroscopic	[ˌsʌbmaɪkrəˈskɒpɪk]	adj. 亚微观的
spinning	[ˈspɪnɪŋ]	n. 纺线；adj. 纺纱的；v. 快速旋转
magnetic	[mæɡˈnetɪk]	adj. 磁的，磁性的
nucleus	[ˈnjuːkliəs]	n. 核心，核子，原子核

Part B　Introduction to Fundamentals of Chemistry and Chemical Engineering　　127

acquisition	[ˌækwɪˈzɪʃ(ə)n]	n. 学得，习得，获取
scanning	[ˈskænɪŋ]	n. 扫描，搜索，观测；adj. 扫描的，观测的
homogeneity	[ˌhəuməudʒəˈniːəti]	n. 同质，同种
superconducting	[ˌsjuːpəkənˈdʌktɪŋ]	adj. 超导的
neodymium	[ˌniːəuˈdɪmiəm]	n. 钕
titanium	[tɪˈteɪniəm]	n. 钛
helium	[ˈhiːliəm]	n. 氦
decane	[ˈdekeɪn]	n. 癸烷
vapor	[ˈveɪpə(r)]	n. 蒸气，水蒸气
chop	[tʃɒp]	v. 砍，剁
bandwidth	[ˈbændwɪdθ]	n. 带宽
nebulizer	[ˈnebjəlaɪzə(r)]	n. 喷雾器
arduous	[ˈɑːdʒuəs]	adj. 艰巨的，艰难的
viscosity	[vɪˈskɒsəti]	n. （液体的）黏性，黏度
calcium	[ˈkælsiəm]	n. 钙
ionization	[ˌaɪənaɪˈzeɪʃn]	n. 离子化
titrimetric	[ˌtaɪtrɪˈmetrɪk]	adj. 滴定的
plasma	[ˈplæzmə]	n. 血浆，等离子（气）体
centrifugation	[sentrɪfjuˈgeɪʃən]	n. 离心分离
branch	[brɑːntʃ]	n. 分支，分支机构

Notes

1. The techniques and methods used and the problems encountered are so diverse that they cut across the traditional boundaries of inorganic, organic, and physical chemistry and include aspects of fields such as biochemistry, physics, engineering, and economics.
所使用的技术和方法以及遇到的问题多种多样，它们跨越了无机化学、有机化学和物理化学的传统划分，并涉及生物化学、物理学、工程学和经济学等领域。

2. Anyone who wants to know more about the composition of substances must use analytical methods to determine the types and amounts of compounds, elements, atoms, and subatomic particles present in a given sample, and to study the detailed compositions of the various species.
任何想要进一步了解物质组成的人，都必须使用分析方法来确定给定样品中存在的化合物、元素、原子和亚原子粒子的类型和数量，并研究各种物质的详细组成。

3. For example, all metallic elements in a sample typically emit ultraviolet and visible radiation when heated to a sufficiently high temperature in an electric arc; all charged species conduct electricity; and all components in a mixture contribute to its refractive index, mass, and volume.
例如，样品中的所有金属元素在电弧中加热到足够高的温度时通常会发出紫外线和可见光辐射；所有带电物质都会导电；混合物中的所有成分都会影响其折射率、质量和体积。

4. Laboratory Information and Management Systems (LIMS) which automatically record large numbers of samples, chemometrics which involves computerized and often sophisticated statistical analysis of data, and expert systems which provide interactive computerized guidance and evaluation in solving analytical problems, have all become important in optimizing chemical analysis and maximizing the information it provides.

   用于自动记录大量样品数据的实验室信息和管理系统（LIMS）、对数据进行计算和复杂的统计分析的化学计量学，以及为解决分析问题提供交互式计算机化指导和评估的专家系统，都已成为优化化学分析和最大限度地利用其所提供信息的重要手段。

5. The interaction of the solute with the mobile and stationary phases can be manipulated by different choices of solvents and stationary phases, giving HPLC a high degree of versatility not found in other chromatographic systems and the ability to easily separate a wide variety of chemical mixtures.

   溶质与流动相和固定相的相互作用可以通过选择不同的溶剂和固定相来控制，这使得高效液相色谱具有其他色谱系统所不具备的高度通用性，能够轻松分离各种化学混合物。

Trivial — Last but not the Least

Qualitative and Quantitative Analysis

Qualitative analysis characterizes the identity of a sample, while quantitative analysis examines its mass or concentration. Techniques used in qualitative analysis include chemical tests, spectroscopy, spectrometry, microscopy, flame tests, and bead tests. Quantitative analysis uses analytical balances, gravimetric analysis, volumetric analysis, and separation techniques such as filtration, centrifugation, and chromatography. There is some overlap in the techniques used between the two branches, especially since samples may need to be purified in order to characterize them.

Unit 8
A Map of Physical Chemistry

8.1 Overview

8.1.1 What Is Physical Chemistry?

Physical chemistry is the branch of chemistry that establishes and develops the principles of the subject, which is an empirical science that studies the physical principles involved in chemical interactions. It studies (1) how matter behaves at the molecular and atomic levels, and (2) how chemical reactions occur.

Physical chemistry is the study of the physical basis of phenomena related to the chemical composition and structure of substances. It has been pursued from two levels, the macroscopic and the molecular. The knowledge of physical chemistry available today provides a rich, comprehensive view of the world of atoms and molecules, linking their nature to the macroscopic properties and phenomena of materials and substances. A starting point for an introduction to physical chemistry is the concept of energy levels in atoms and molecules, distributions among these energy levels, and something familiar, temperature.

Physical chemistry is the study of the underlying physical principles that govern the properties and behavior of chemical systems, based on four major theoretical areas: thermodynamics, kinetics (or, more generally, transport processes), quantum mechanics, and statistical mechanics[1]. Physical chemistry is a fascinating subject. It is fair to say that many parts of physics and all parts of chemistry are included in physical chemistry and its applications. Furthermore, it is the course in which most chemistry students first have the opportunity to synthesize what they have learned in mathematics, physics, and chemistry courses into a coherent pattern of knowledge. We see it as the quantitative interpretation of the macroscopic world in terms of the atomic-molecular world. To achieve this interpretation, we must organize our observations of macroscopic phenomena, as we do in thermodynamics and parts of kinetics. We must advance our studies of atoms and molecules, as we do, for example, in quantum mechanics and spectroscopy. Then we have to bring these studies together. This coming together is woven into much of the fabric of a modern physical chemistry course.

Physical chemistry is the application of the methods of physics to chemical problems. It involves the qualitative and quantitative study, both experimental and theoretical, of the general principles governing the behavior of matter, especially the transformation of one substance into another. Although the physical chemist uses many of the methods of the physicist, he applies them to chemical structures and chemical processes. Physical chemistry is not so much concerned with the description of chemical substances and their reactions, which is the concern of organic and inorganic chemistry, as it is with theoretical principles and quantitative problems.

Research by physical chemists is an increasingly small part of industrial research. Accordingly, fewer physical chemists are being hired by industry and government laboratories. Nevertheless, physical chemistry provides a broad education and positions students to work in a variety of scientific careers, such as (1) emerging fields in materials science and molecular modeling. Combining the traditional mathematical rigor of physical chemistry with the practicality of these fields offers new and exciting opportunities, and (2) careers in analytical chemistry. Here, you'll work to understand the fundamental processes involved in analytical techniques and look for ways to improve and extend them.

8.1.2 Theories of Physical Chemistry

Traditionally, there are three main areas of physical chemistry: thermodynamics (which deals with the energetics of chemical reactions), quantum chemistry (which deals with the structures of molecules), and chemical kinetics (which deals with the rates of chemical reactions). The theoretical foundations of these subjects are, respectively, quantum mechanics, thermodynamics and equilibrium statistical mechanics, and chemical kinetics and kinetic theory. These theories, firmly grounded in experimental findings, provide the structure necessary to understand past achievements and to recognize and develop important new areas of physical chemistry.

Thermodynamics, as applied to chemical problems, is primarily concerned with the location of chemical equilibrium, the direction of chemical change, and the associated changes in energy. Quantum chemistry theoretically describes bonding at the molecular level and, in its exact treatments, deals only with the simplest of atomic and molecular systems, but it can be extended in an approximate way to deal with bonding in much more complex molecular structures. Chemical kinetics is concerned with the rates and mechanisms by which processes occur as equilibrium is approached.

An intermediate area, known as statistical thermodynamics, encompasses the three main areas of thermodynamics, quantum chemistry, and kinetics, and also provides a fundamental relationship between the microscopic and macroscopic worlds. Related to this is non-equilibrium statistical mechanics, which is becoming an increasingly important part of modern physical chemistry. This area includes problems in areas such as the theory of fluid dynamics and light scattering.

Two approaches are possible in a physicochemical study. In what might be called a systemic approach, the investigation begins with the very basic constituents of matter, the fundamental particles, and proceeds conceptually to construct larger systems from them. The adjective microscopic is used to refer to these tiny constituents. In this way, increasingly complex phenomena can be interpreted on the basis of elementary particles and their interactions. The second approach begins with the study of macroscopic material, such as a sample of a liquid or solid that can be easily observed with the naked eye. Measurements are made of macroscopic properties such as pressure, temperature, and volume. In the phenomenological approach, more detailed studies of microscopic behavior are made only as necessary to understand the macroscopic behavior in terms of the microscopic[2].

The topics of traditional physical chemistry can be grouped into several areas: (1) the study of the macroscopic properties of systems of many atoms or molecules; (2) the study of the processes that systems of many atoms or molecules can undergo; (3) the study of the properties of individual atoms and molecules; and (4) the study of the relationship between molecular and macroscopic properties.

Physical chemistry is a good field for chemists who have a strong curiosity about how things work at the atomic level and who enjoy working with laboratory equipment and machinery.

8.1.3 What Do Physical Chemists Do?

Physical chemists discover, test, and seek to understand the physical properties of a material (i.e., solid, liquid, or gas). Precision and attention to detail make their work similar to analytical chemistry. They use sophisticated instruments and equipment such as lasers, mass spectrometers, nuclear magnetic resonance, and electron microscopes to (1) analyze materials, (2) develop methods for testing and characterizing the properties of materials, (3) develop theories about these properties, and (4) discover potential uses for materials.

Physical chemists focus on understanding the physical properties of atoms and molecules, how chemical reactions work, and what these properties reveal. Their discoveries are based on understanding chemical properties and describing their behavior using theories of physics and mathematical calculations. Physical chemists emphasize the importance of applying mathematics to their work. They use mathematical analysis and statistics on large data sets—sometimes with millions of data points—to reveal hidden information about compounds, materials, and processes. They may also run simulations, developing mathematical equations that predict how compounds will react over time. Many who work in the lab say their time is split between the bench and the desk, where they perform calculations and review data. Physical chemists who move into management also spend time supervising other scientists, reviewing departmental needs and goals, and meeting with business managers in their companies.

8.1.4 Perspective of Physical Chemistry

Physical chemistry provides the theoretical foundation for all of chemistry and many related subjects. Therefore, I believe, as do many instructors, that the first course in physical chemistry should lead to a critical understanding of the primary theoretical concepts and their use in explaining important experiments. The universal lessons of physical chemistry are quantitative reasoning, problem solving, rigorous and accurate thinking. Many students may never directly use the factual knowledge gained in a course in physical chemistry, but all can benefit from the skills and habits learned. In the opinion of some, this emphasis leads to excessive detail in certain places. But attention to detail is the essence of good science, and in any case it is easier to ignore unnecessary details than to add important ones that are missing.

Your chemistry education has trained you to think in terms of molecules and their interactions, and we believe that a course in physical chemistry should reflect this viewpoint. The focus of modern physical chemistry is the molecule. Current experimental research in physical chemistry uses equipment such as molecular beam machines to study the molecular details of gas-phase chemical reactions, high vacuum machines to study the structure and reactivity of molecules at solid interfaces, lasers to determine the structure of individual molecules and the dynamics of chemical reactions, and nuclear magnetic resonance spectrometers to learn about the structure and dynamics of molecules[3]. Modern theoretical research in physical chemistry uses the tools of classical mechanics, quantum mechanics, and statistical mechanics along with computers to develop a detailed understanding of chemical phenomena in terms of the structure and dynamics of the molecules involved. For example, computer calculations of the electronic structure of molecules provide fundamental insights into chemical bonding, and computer simulations of the dynamic interaction between molecules and proteins are used to understand how proteins function.

Physical chemists are inordinately proud of "back of the envelope" calculations; napkins and paper placemats have also proven useful. Everyone makes mistakes, but smart people find those mistakes before they do damage. Chemistry deals with an enormous number of substances; it is a vast science. Physics, on the other hand, deals with rather few substances; it is an intensive science. Physical chemistry is the child of these two sciences; it has inherited the comprehensive character of chemistry. This is the reason why its comprehensive character, which has attracted so much admiration. Physical chemistry may be regarded as an excellent school of exact reasoning for all students of the natural sciences.

8.2 Briefing on Thermodynamics

Thermodynamics is the branch of physics that deals with the relationship between heat and other properties (such as pressure, density, temperature, etc.) in a substance. Specifically, thermodynamics is largely concerned with how heat transfer is related to various energy changes within a physical system undergoing a thermodynamic process. Such processes usually result in work being done by the system and are governed by the laws of thermodynamics.

The first law of thermodynamics is the law of conservation of energy. This law states that energy cannot be created or destroyed. The second law allows the prediction of the direction of change. According to the second law, the entropy of a system may decrease, but the entropy of the universe increases. Finally, the third law states that the entropy of perfect crystalline substances is zero at absolute zero, which is unattainable.

The bond energy refers to the average energy requirement for breaking a covalent bond in a polyatomic molecule. These energies can be applied to estimate $\Delta H_r^{\ominus}$ for reactions of reactive species that cannot be isolated, or the standard enthalpies of formation for compounds that have not yet been prepared.

Entropy is a measure of the disorder of a system and the probability of a system being in a particular state. The more arrangements that are possible for the molecules of a sample, or the more energy levels that can be occupied, the greater the entropy. Entropy increases with temperature. Although all spontaneous changes increase the entropy of the universe, the entropy of a system can either increase or decrease with a spontaneous change. If the entropy of the system decreases, the entropy of the environment must increase more. Standard entropy, $S^{\ominus}$, is the entropy of a substance in a standard state. Entropy is a state function; calculating standard entropy changes from standard entropies is similar to calculating standard enthalpy changes from standard enthalpies of formation.

The free energy, G, combines both enthalpy and entropy into a singular state function. The equation that establishes the connection between the standard changes in free energy, enthalpy, and entropy changes is: $\Delta G^\ominus = \Delta H^\ominus - T\Delta S^\ominus$

Standard free energy changes can be determined either by the definition of standard free energy changes or by using standard free energy changes of formation. For changes made under non-standard conditions, the free energy changes can be calculated using the equation: $\Delta G = \Delta G^\ominus + RT \ln Q$

If spontaneous change is to occur, ΔG must be negative. The spontaneous direction of the change is determined by the temperature when ΔH and ΔS have the same sign. The definition of free energy change can be used to estimate free energy changes at different temperatures if ΔH and ΔS do not change with temperature.

The $\Delta G^\ominus$ is related to the equilibrium constant by the equation: $\Delta G^\ominus = -RT \ln K$

This equation can assist in determining the value of K or in calculating $\Delta G^\ominus$ if K is already known. The maximum amount of energy available to perform useful work is represented by the free energy lost by the system during a spontaneous change at constant temperature and pressure. The minimum work required to cause a nonspontaneous change at constant temperature is represented by the free energy gained by the system.

Thermodynamics tells us whether a change can occur spontaneously and how far a change will go before reaching equilibrium. It does not tell you how fast a change will occur or how the change will occur.

8.3 Basic Concepts of Thermodynamics

(1) Heat transfer

Broadly speaking, heat in a material is recognized as measuring the energy contained within the particles of that material. This concept is referred to as the kinetic theory of gases, although it can be applied to solids and liquids to varying degrees. The heat generated from the motion of the particles can be transferred to neighboring particles, and thus to different parts of the material or other materials, in a variety of ways:

✧ Thermal contact occurs when two substances can affect each other's temperature

✧ Thermal equilibrium occurs when two substances in thermal contact stop transferring heat

✧ Thermal expansion occurs when a substance increases in volume as it gains heat. Thermal contraction also occurs

✧ Conduction occurs when heat flows through a heated solid

✧ Convection occurs when heated particles transfer heat to another substance, such as cooking something in boiling water

✧ Radiation is when heat is transferred by electromagnetic waves, such as from the sun

✧ Insulation is when a poorly conducting material is used to prevent heat transfer

(2) Thermodynamic processes

A system undergoes a thermodynamic process when there is some sort of energetic change within the system, generally associated with changes in pressure, volume, internal energy (i.e., temperature), or some sort of heat transfer. There are several specific types of thermodynamic processes that have particular characteristics:

✧ Adiabatic process — a process in which no heat is transferred into or out of the system.

✧ Isochoric process — a process with no change in volume, in which case the system does no work.

✧ Isobaric process — a process in which there is no change in pressure.

✧ Isothermal process — a process in which there is no change in temperature.

(3) States of matter

A state of matter is a description of the kind of physical structure that a material substance manifests, with properties that describe how the material holds together. There are five states of matter, although only the first three are usually included in the way we think about states of matter:

✧ Gas
✧ Liquid
✧ Solid
✧ Plasma
✧ Superfluid (such as a Bose-Einstein condensate)

Many substances can transition between the gas, liquid, and solid phases of matter, while only a few rare substances are known to be able to enter a superfluid state. Plasma is a special state of matter, such as lightning.

✧ Condensation — gas to liquid
✧ Freezing — liquid to solid
✧ Melting — solid to liquid
✧ Sublimation — solid to gas
✧ Vaporization — liquid or solid to gas

(4) Heat capacity

The heat capacity, C, of an object is the ratio of the change in heat (energy change, ΔQ, where the Greek symbol delta, Δ, denotes a change in quantity) to the change in temperature (ΔT): $C = \Delta Q / \Delta T$

The heat capacity of a substance indicates the ease with which a substance heats up. A good thermal conductor would have a low heat capacity, indicating that a small amount of energy causes a large temperature change. A good thermal insulator would have a large heat capacity, indicating that a large amount of energy transfer is required for a temperature change.

(5) Ideal gas equations

There are several ideal gas equations that relate temperature (T_1), pressure (P_1), and volume (V_1). These values after a thermodynamic change are indicated by (T_2), (P_2), and (V_2). For a given amount of a substance, n (measured in moles), the following relationships apply

✧ Boyle's Law (T is constant):

$$P_1 V_1 = P_2 V_2$$

✧ Charles/Gay-Lussac's Law (P is constant):

$$V_1/T_1 = V_2/T_2$$

✧ Ideal Gas Law:

$$P_1 V_1/T_1 = P_2 V_2/T_2 = nR$$

R is the ideal gas constant, $R = 8.3145$ J/(mol·K). Therefore, for a given amount of matter, nR is constant, which gives the ideal gas law.

(6) Laws of thermodynamics

✧ Zeroth Law of Thermodynamics — When three systems are in thermal equilibrium, they are in thermal equilibrium with one another

✧ First Law of Thermodynamics — The change in energy of a system is equivalent to the energy added to or removed from the system

✧ Second Law of Thermodynamics — It is impossible for a process to solely result in the transfer of heat from a cooler body to a hotter one

✧ Third Law of Thermodynamics — It is impossible to bring any system to absolute zero in a finite number of operations. Consequently, a perfectly efficient heat engine cannot be constructed

(7) The second law & entropy

The second law of thermodynamics can be restated in terms of entropy, which quantifies disorder in a system. Entropy change of the process is the change in heat divided by the absolute temperature. Following this definition, the second law can be rephrased: in any closed system, the entropy of the system will either remain constant or increase. By "closed system", we indicate that every part of the process gets counted in the system's entropy calculation.

(8) More about thermodynamics

In some ways, thermodynamics cannot be considered a distinct discipline of physics. Since energy transformation is present in nearly every aspect of physics, from astrophysics to biophysics, thermodynamics plays an important role. Without the ability of systems to utilize energy to perform work — the core principle of thermodynamics — physicists would have no subject to contemplate.

However, there are some fields that use thermodynamics in passing as they go about studying other phenomena, while there are a wide range of fields that focus heavily on the thermodynamic situations involved. Here are some of the subfields of thermodynamics:

✧ Cryophysics / Cryogenics / Low Temperature Physics — the study of physical properties in extremely cold environments that are far below the temperatures found in even the coldest regions of the Earth. Superfluids are an example of a phenomenon studied in this field.

✧ Fluid Dynamics — the study of the physical properties of "fluids", in this case specifically defined as liquids and gases.

✧ High Pressure Physics — the study of the physics of extremely high-pressure systems, generally related to fluid dynamics.

✧ Meteorology / Weather Physics — the physics of weather, atmospheric pressure systems, etc.

✧ Plasma Physics — the study of matter in a plasma state.

8.4 Introduction to Chemical Equilibrium and Kinetic Theory

8.4.1 Equilibrium

A major goal of chemists is to understand chemical reactions—to know whether, under a given set of conditions, two substances will react when mixed, to determine whether a given reaction will be exothermic or endothermic, and to predict how far a given reaction will proceed before equilibrium is reached[4]. An equilibrium state, produced as a result of two opposing reactions occurring simultaneously, is a state in which there is no net change as long as the conditions do not change. This section shows how to predict the equilibrium states of chemical systems using thermodynamic data and how experimental measurements of equilibrium states can provide valuable thermodynamic information. However, it should be noted that thermodynamics alone is insufficient in explaining the rate at which equilibrium is established or providing specific details about the mechanism by which equilibrium is reached. To obtain such explanations, the quantum theory of molecular structure and statistical mechanics must be considered.

To fully understand the nature of the chemical equilibrium state, it is first necessary to have some familiarity with the factors that affect reaction rates. The factors that affect the rates of a chemical reaction are temperature, concentrations of reactants (or partial pressures of gaseous reactants), and the presence of a catalyst. In general, for a given reaction, the higher the temperature, the faster the reaction. The concentrations of reactants or partial pressures of gaseous reactants affect the rate of the reaction; an increase in concentration or partial pressure will increase the rate of most reactions. Substances that speed up a chemical reaction, but are not themselves consumed in the reaction are called **catalysts**.

Dynamic equilibrium

In many cases, direct reactions between two substances seem to cease before all of the starting materials are exhausted. In addition, the products of chemical reactions themselves often react themselves to produce the starting materials, for example, nitrogen and hydrogen combine at 500°C in the presence of a catalyst to produce ammonia:

$$N_2 + 3H_2 \longrightarrow 2NH_3$$

At the same temperature and in the presence of the same catalyst, pure ammonia decomposes into nitrogen and hydrogen:

$$2NH_3 \longrightarrow 3H_2 + N_2$$

For convenience, these two opposite reactions are represented in the same equation by a double arrow:

$$N_2 + 3H_2 \rightleftharpoons 2NH_3$$

The reaction proceeding toward the right is called the forward reaction; the other is called the reverse reaction.

When ammonia or a mixture of nitrogen and hydrogen is subjected to the above conditions, a mixture of all three gases is produced. The rate of reaction between the materials added to the reaction vessel will decrease after the reaction begins because their concentrations will decrease. Conversely, the material being produced will react faster after the reaction starts because there will be more of it. Thus, the faster forward reaction slows down, and the slower backward reaction speeds up. Eventually, a point is reached where the rates of the forward and reverse reactions are balanced and no more net changes occur, this is known as equilibrium. Equilibrium is a dynamic state because both reactions are still proceeding, but since the two opposite reactions are going on at equal rates, no net change is observed.

All chemical reactions eventually proceed toward equilibrium. In practice, however, some reactions go so far in one direction that the reverse reaction cannot be detected, and they are said to proceed to completion. The principles of chemical equilibrium apply even to these, and it will be seen that for many of them the extent of the reaction can be expressed quantitatively.

Equilibrium constants

Equilibrium is a state of dynamic equilibrium between two opposing processes. For a common reactant at a given temperature, we have:

$$A + B \rightleftharpoons C + D$$

At the point of equilibrium, the following ratio must be a constant:

$$K = \frac{[C][D]}{[A][B]}$$

The constant, K, is called the equilibrium constant of the reaction. It has a specific value at a given temperature. If the concentration of any component in equilibrium system is changed, the concentrations of the other components will change as well so that the specified ratio remains equal to K, assuming the temperature stays constant. The equilibrium constant expression quantitatively defines the equilibrium state.

More generally, for the reversible reaction:

$$a A + b B \rightleftharpoons c C + d D$$

The equilibrium constant expression is written as follow:

$$K = \frac{[C]^c[D]^d}{[A]^a[B]^b}$$

The equilibrium constant expression conventionally places the concentration terms of reaction products in the numerator. It's important to note that the exponents of the concentration terms in the expression correspond to the coefficients of the respective species in the balanced chemical equation.

In typical acid-base neutralization reactions, the relative positions of the acid and base in a table of K_a values can be utilized to predict the direction of the equilibrium between the two components. The stronger acid and the stronger base react to form the weaker acid and the weaker base. The value of the equilibrium constant can be determined by using the values of the equilibrium constants for the two weak acids and applying the principle that the equilibrium constant for the sum of two or more equations is the product of the equilibrium constants for the initial equations.

The strength of acids and bases in water solutions is determined by several factors, the most important of which are the polarity and strength of the chemical bonds, as well as the degree of solvation. Fortunately, some empirical rules relating acidity to structure work relatively well at room temperature. The acidity of the binary compounds (compounds consisting of two elements) of hydrogen increases from left to right when moving across rows and from top to bottom within groups of the periodic table. The hydrides of the reactive metals of Groups ⅠA and ⅡA exhibit strong basic properties, while those of several nonmetals exhibit acidic properties.

The hydroxides of the reactive metals act as strong bases, while the hydroxides of the nonmetals exhibit acidic behavior. Elements close to the metal-nonmetal transition display amphoteric properties. The acidity of oxo acids is influenced by the electronegativity of the central atom and the number of additional oxygens attached to it. The acidity of organic acids is increased by substituting more electronegative atoms for hydrogens near the —COOH group.

Hydrated cations also act as Brønsted-Lowry acids. When the charge and radius are similar, the electronegativity of the metal determines the acidic acidity of the hydrated cation. As they move down the periodic table, the hydrated cations become less acidic. The higher the charge, when the radii and electronegativities are similar, the more acidic the hydrated cation. The acidity of the hydrated cation increases as the radius of the ion decreases while the charge and electronegativity remain similar.

The basicity of anionic and cationic weak bases can be deduced from the acidity of their conjugate acids. Anions containing one or more ionizable hydrogens, such as HCO^-, are amphoteric, as are species such as $Al(H_2O)_5(OH)^{2+}$.

Water balances the acidity of strong acids and bases. A solvent that shows differences in acidity between acids or differences in basicity between bases is called a **differentiating solvent**. Most metal oxides are alkaline. Oxides of some metals near the metal-nonmetal transition are amphoteric. Most nonmetallic oxides are acidic, and many nonmetallic oxides react with water to form oxo acids. When an element forms more than one oxide, the oxide with the higher oxygen content reacts with water to form a stronger acid. A few nonmetallic oxides are neither acidic nor basic. Acidic and basic oxides often react directly to form salts.

Lewis acids are not limited to hydrogen containing compounds. A Lewis acid is a species that can attach to an electron pair to form a covalent bond and requires an empty valence orbital for the electron pair to enter. Species with incomplete octets, small cations with high charges, and species with multiple bonds are Lewis acids. Conversely, a Lewis base is a species with an unshared electron pair that can form a covalent bond, and it must possess an unshared valence electron pair. Lewis acid-base reactions are essentially substitution reactions in which one Lewis base replaces another.

Complexes are species in which a central atom is surrounded by covalently bonded Lewis bases, which are commonly referred to as **ligands**. Complexes may be in the form of ions or molecules. The formation of complexes other than aqua complexes in aqueous solution involves the substitution of water by other ligands: the substitution of water by other ligands is gradual and reversible. The equilibrium constants for each step decrease, yet the differences between them are typically small, resulting in a mixture of complexes in the solution. Complex ion equilibria are generally too complicated to be quantitatively analyzed in general chemistry, unless an excess of ligand results in a dominant single species.

The solubility product constant, or K_{sp}, is equal to the concentration product of ions in a saturated solution of a slightly soluble salt, with each raised to the power of its coefficient in the equilibrium equation. The solubility of a few slightly soluble salts and hydroxides is quantitatively related to the equilibrium constant. For most, however, the actual solubilities are higher than the solubilities calculated from K_{sp}. For some, the measured solubility is significantly higher than the calculated solubility. Solubility equilibria are affected by various factors such as the common ion effect, the salt effect, and side reactions like hydrolysis, incomplete dissociation, and ion pair formation. A salt effect occurs when the solubility of a compound is increased by the presence of non-common ions in the solution.

Some poorly soluble salts and metal hydroxides can be dissolved by forming a complex ion. Others can be dissolved by Brønsted-Lowry acid-base reactions.

8.4.2 Chemical Kinetics

Chemical kinetics studies the rates of chemical reactions, providing important source of information about reaction mechanisms and how they occur. Mechanisms are theoretical and can never be proven correct, although they can be shown to be wrong. They are useful for organizing information about chemical reactions. The rate of a reaction can be determined by tracking the amounts or concentrations of reactants or products at intervals because the rates of formation and disappearance are linked by the reaction equation. Several factors affect the rate of a chemical reaction, including the identities and concentrations of the reactants, temperature, the presence of catalysts, solvents, and, in the case of heterogeneous reactions, particle size and mixing.

When a chemical system reaches equilibrium, the forward and reverse reactions occur at the same rate. It is important to determine the rate at which reactants are consumed or products are formed in a process. It is also critical to have comprehensive data on reactant rates in order to evaluate theories and mechanisms for various chemical processes. Experimental results indicate that several reaction variables significantly affect reaction rates.

◆ **Temperature.** Temperature has a significant effect on the rate of chemical reactions. Therefore, a typical laboratory procedure involves performing reactions at a constant temperature (isothermal) to eliminate one variable while studying the reaction kinetics.

◆ **Pressure and volume.** Volume and pressure are important in the kinetic study of gas phase reactions. Usually, the volume is fixed by running the reaction in a vessel of fixed dimensions. For solid and liquid phase reactions, the pressure is usually atmospheric, and the volume of the reacting system is relatively unimportant because there is little change in volume.

◆ **Concentration.** At any given temperature, the rates of most chemical reactions are functions of the concentrations of one or more components of the system. In practice, the concentrations of the reactants are usually used to determine the overall rates of the reactions.

◆ **Catalyst.** Any substance that affects the rate of a chemical reaction but cannot be identified as a product or reactant is called a catalyst. A catalyst can accelerate or decelerate the rate, but we usually refer to decelerating catalysts as inhibitors.

The dependence of the reaction rate on the concentration is described by the rate law of the reaction, which must be determined by experiment. Many rate equations are expressed as rate = $k[A]^x[B]^y$, where k is a constant known as the rate constant. The value of the rate constant depends on all factors that affect the rate of the reaction except the concentration. When the exponent of a concentration in the rate law is zero, the reaction is of zero order with respect to that substance. As a result, the rate of the reaction becomes independent of the concentration of that substance. If the exponent is one, the reaction follows a first order relationship with respect to the substance, resulting in a rate that doubles when the concentration of that reactant is doubled. If the exponent is two, the reaction follows a second order relationship with respect to the substance, resulting in a rate that increases fourfold when the concentration of that substance is doubled. The overall order of the reaction is determined by adding the orders with respect to each reactant.

The rate law may be determined by observing or graphing the initial rates, while the half-life of a reaction, $t_{1/2}$, represents the time required for half of the initial amount of reactant to react.

The rates of almost all reactions increase with increasing temperature. The Arrhenius equation, $k = Ae^{-Ea/RT}$ quantifies this relationship. The collision and activated complex (transition state) theories explain the dependence of reaction rates on concentration and temperature.

The collision theory states that chemical reactions occur through collisions between molecules or ions. According to the collision theory, particles must collide with enough energy, specifically the activation energy, and in the proper orientation. According to the activated complex theory, the reactants follow the reaction coordinate, which is the path of the least available energy, from reactants to activated complex to products. The activated complex represents the arrangement of atoms with the highest energy along the reaction coordinate. Reaction profiles are graphical representations that depict the changes in potential energy observed as the reactants move along the reaction coordinate from reactants to activated complex to product.

Most reaction mechanisms consist of a series of simple steps called elementary processes. The molecularity of an elementary process is the number of particles that must collide to form the activated complex for the elementary process. In the case of elementary processes, the rate law exponents correspond to the coefficients stated in the chemical equation. If the exponents in the overall rate law for the reaction differ from the coefficients in the equation, this indicates that the mechanism likely involves multiple steps. On the other hand, if the exponents in the rate law match the coefficients in the equation, the reaction may or may not occur in a single step. It is critical that the steps in any reaction mechanism sum up to a net equation for the overall reaction. Furthermore, the mechanism must be consistent with the experimental evidence from the reaction.

Intermediates are chemical species that are formed in one step and consumed in another. At a steady state, the concentration of an intermediate remains constant because it is formed in one step and consumed in a later step. A rate-determining step is a step in the sequence of elementary processes in a reaction mechanism that is much slower than the other steps and determines the rate of the overall reaction.

Catalysts provide an alternative pathway with a lower activation energy and/or increase the collision frequency in the correct orientation. Catalysts may be homogeneous or heterogeneous. Catalysts are present at both the beginning and the end of a reaction; they are consumed in one step and reformed in another. According to the principle of microscopic reversibility, the reverse reaction follows the same path as the forward reaction, but in the opposite direction.

Enzymes are complex substances produced in living cells that catalyze biological reactions. The reactant in an enzyme-catalyzed reaction is called the substrate. The substrate fits into the active site of the enzyme.

New Words and Expressions

branch	[brɑːntʃ]	n. 树枝，分枝，分支机构；v. 分岔，岔开
macroscopic	[ˌmækrəʊˈskɒpɪk]	adj. 宏观的，肉眼可见的
thermodynamics	[ˌθɜːməʊdaɪˈnæmɪks]	n. 热力学

kinetics	[kaɪˈnetɪks]	n. 反应动力学，动力学
quantum	[ˈkwɒntəm]	n. 量子，量子论；adj. 量子的
theoretical	[ˌθɪəˈretɪk(ə)l]	adj. 理论的，理论上存在的
equilibrium	[ˌiːkwɪˈlɪbriəm]	n. 平衡，均衡
dynamics	[daɪˈnæmɪks]	n. 动力学，力学，动力
scattering	[ˈskætərɪŋ]	n. 散射，分散；adj. 分散的
curiosity	[ˌkjʊəriˈɒsəti]	n. 好奇心，求知欲
laser	[ˈleɪzə(r)]	n. 激光（器）
spectrometer	[spekˈtrɒmɪtə(r)]	n. 分光仪
resonance	[ˈrezənəns]	n. 洪亮，共鸣，共振
hidden	[ˈhɪd(ə)n]	adj. 隐形的，隐秘的
rigorous	[ˈrɪɡərəs]	adj. 严密的，严格的
vacuum	[ˈvækjuːm]	n. 真空，空白
napkin	[ˈnæpkɪn]	n. 餐巾纸
inherit	[ɪnˈherɪt]	v. 继承，继任
entropy	[ˈentrəpi]	n. 熵
polyatomic	[ˌpɒliəˈtɒmɪk]	adj. 多原子的
enthalpy	[enˈθælpɪ]	n. 焓，热焓
spontaneous	[spɒnˈteɪniəs]	adj. 自发的
insulation	[ˌɪnsjuˈleɪʃ(ə)n]	n. 隔热，绝缘，隔音
adiabatic	[ˌædɪəˈbætɪk]	adj. 绝热的，隔热的
isochoric	[aɪsəˈkɒrɪk]	adj. 等容的
isobaric	[ˌaɪsəʊˈbærɪk]	adj. 等压的，恒压的
isothermal	[ˌaɪsəʊˈθɜːməl]	adj. 等温的，等温线的；n. 等温线
superfluid	[ˈsjuːpəˌfluːɪd]	n. 超流体；adj. 超流体性的，超流体的
minus	[ˈmaɪnəs]	prep. 减，减去，缺少，零下；n. 减号，负号；adj. 不利的，负面的
engine	[ˈendʒɪn]	n. 引擎，驱动力，推动力
virtually	[ˈvɜːtʃuəli]	adv. 事实上，几乎
subfield	[ˈsʌbfiːld]	n. 子域，分区
region	[ˈriːdʒən]	n. 地区，区域
exothermic	[ˌeksəʊˈθɜːmɪk]	adj. 发热的，放热的
endothermic	[ˌendəʊˈθɜːmɪk]	adj. 吸热的，温血的
cease	[siːs]	v. 终止，结束；n. 停止
reverse	[rɪˈvɜːs]	v. 逆转，撤销；n. 相对，相反
reactant	[riˈæktənt]	n. 反应物，反应剂
reversible	[rɪˈvɜːsəb(ə)l]	adj. 可逆的
numerator	[ˈnjuːməreɪtə(r)]	n. 分子，计算者，计算器
coefficient	[ˌkəʊɪˈfɪʃ(ə)nt]	n. 系数
neutralization	[ˌnjuːtrəlaɪˈzeɪʃ(ə)n]	n. 中和，中和作用
polarity	[pəˈlærəti]	n. 极性，两极，对立
binary	[ˈbaɪnəri]	n. 二进制

bottom	[ˈbɒtəm]	n. 底，底部，尽头；adj. 底部的，最后的
periodic	[ˌpɪəriˈɒdɪk]	adj. 定期的，元素周期表的
amphoteric	[ˌæmfəˈterɪk]	adj. 两性的
acidity	[əˈsɪdəti]	n. 酸度，酸性
radius	[ˈreɪdiəs]	n. 半径，半径范围，周围
anionic	[ænaɪˈɒnɪk]	adj. 阴离子的
cationic	[ˌkætaɪˈɒnɪk]	adj. 阳离子的；n. 阳离子
ionizable	[ˈaɪənaɪzəbl]	adj. 可电离的
valence	[ˈveɪləns]	n. 价，化合价
orbital	[ˈɔːbɪt(ə)l]	adj. 轨道的
octet	[ɒkˈtet]	n. 八隅体
ligand	[ˈlɪgənd]	n. 配位体，配基
aqueous	[ˈeɪkwiəs]	adj. 水的，水般的
gradual	[ˈgrædʒuəl]	adj. 逐渐的，渐进的
predominate	[prɪˈdɒmɪneɪt]	v. 占优势，占主导地位
vessel	[ˈves(ə)l]	n. 船，舰，血管，导管，容器
interval	[ˈɪntəvl]	n. 间隔，（时间上的）间隙，间歇
decelerate	[ˌdiːˈseləreɪt]	v. 减速，降低速度
exponent	[ɪkˈspəʊnənt]	n. 指数，幂
collision	[kəˈlɪʒ(ə)n]	n. 碰撞，相撞，冲突
orientation	[ˌɔːriənˈteɪʃ(ə)n]	n. 目标，定位，方向，朝向
molecularity	[məʊlekjuˈlærɪti]	n. 分子性，反应分子数
elementary	[ˌelɪˈmentri]	adj. 基本的，基础的，初级的
homogeneous	[ˌhɒməˈdʒiːniəs]	adj. 同种类的，同性质的
heterogenous	[ˌhetəˈrɒdʒənəs]	adj. 异质的，异种的
enzyme	[ˈenzaɪm]	n. 酶

Notes

1. Physical chemistry is the study of the underlying physical principles that govern the properties and behavior of chemical systems, based on four major theoretical areas: thermodynamics, kinetics (or, more generally, transport processes), quantum mechanics, and statistical mechanics.
   物理化学是一门研究支配化学系统特性和行为的基本物理原理的学科，它以四大理论领域为基础：热力学、动力学（或更广义的传输过程）、量子力学和统计力学。

2. Measurements are made of macroscopic properties such as pressure, temperature, and volume. In the phenomenological approach, more detailed studies of microscopic behavior are made only as necessary to understand the macroscopic behavior in terms of the microscopic.
   对压力、温度和体积等宏观属性进行测量。在现象学方法中，只有在必要时才会对微观行为进行详细的研究，以便从微观角度理解宏观行为。

3. Current experimental research in physical chemistry uses equipment such as molecular

beam machines to study the molecular details of gas-phase chemical reactions, high vacuum machines to study the structure and reactivity of molecules at solid interfaces, lasers to determine the structure of individual molecules and the dynamics of chemical reactions, and nuclear magnetic resonance spectrometers to learn about the structure and dynamics of molecules.

目前的物理化学实验研究使用的设备包括：研究气相化学反应分子细节的分子束机、研究固体界面分子结构和反应性的高真空机、确定单个分子结构和化学反应动力学的激光器，以及了解分子结构和动力学的核磁共振光谱仪。

4. A major goal of chemists is to understand chemical reactions—to know whether, under a given set of conditions, two substances will react when mixed, to determine whether a given reaction will be exothermic or endothermic, and to predict how far a given reaction will proceed before equilibrium is reached.

化学家的一个主要目标是了解化学反应，知道在给定条件下，两种物质混合后是否会发生反应，确定给定反应是放热还是吸热，以及预测给定反应在达到平衡之前会进行到什么程度。

Trivial — Last but not the Least

SI Unit Rules and Style Conventions

计量单位（Unit of Measurement）是科技英语的重要组成部分，在阅读或书写科技论文时，经常会遇到现在已较普遍应用的国际单位制（International System of Units，简写 SI），也会遇到非 SI 单位。因此，了解 SI 及其有关单位，对准确理解原文和正确使用它都是很有必要的。

1. SI 的三类单位

(1) Basic units

m (length)	meter（长度）米	K (thermodynamic temperature)	kelvin（热力学温度）开尔文
kg (mass)	kilogram（质量）千克	mol (amount of substance)	mole（物质的量）摩尔
A (electric current)	ampere（电流）安培	cd (luminous intensity)	candela（光强度）坎德拉

(2) Derived units

◇ 用基本单位和辅助单位的代数式表达，包括空间和时间，周期及有关现象，如力、热、光、电、声、物理、化学、核反应等量的单位。例如：m/s (velocity) meter per second（速度）米每秒；$kg \cdot m^2$ (moment of inertia) kilogram square meter（转动惯量）千克平方米

◇ 除上述表述外，有些单位还有专门名称符号（多为科学家名字命名，共 15 个）。例如：

N ($kg \cdot m/s^2$) (force)   newton（力）牛顿
J ($kg \cdot m^2/s$) (work)   joule（功）焦耳

(3) Supplementary units

rad (plane angle)   radian（平面角）弧度
sr (solid angle)   steradian（立体角）球面度

2. SI 单位使用规则

◇ 单位符号、词头符号、运算符号、缩写符号和特殊函数符号用正体（Roman type）而不用斜体（Italics）书写；表示量和物理量的字母如变量、矩阵和坐标轴用斜体书写

◇ 单位符号无复数。例：15m <u>不应</u>是 15ms

◇ 在量和单位的符号后面不能附加圆点（正常的句子末尾的句号除外），因为它们不是缩略语。例如 15m **不应**为 15m．

◇ 除了来源于人名的单位符号第一个字母要大写（如 Pa、Hz、Wb 等）外，其余均为小写字母；但升的符号"L"除外，科学家名字做单位写全称时也用小写；例如 Pa（pascal），N（newton），rad（弧度），s（秒）等；词头符号如：纳（n），微（μ），千（k）用小写；常见有 nm（纳米），kN（千牛），kHz（千赫兹）

◇ 使用 SI 单位时，符号不应随意改变，例：s（秒）不应写作 sec，m（米）不应写作 M

◇ 两个或两个以上单位可用下面（乘或除）几种形式构成组合单位；若非表示两个单位相乘，在符号间不得留空隙，更不能夹杂有其他符号或数字；表示"除"的形式中，斜线在同一行内不得多于一条。例如：N·m 牛·米；m/s（m·s^{-1}）米 / 秒

◇ 在表示量值的和或差范围时，应分别写出单位或加括号将数值组合。如：(25±2)℃ 不应写作 25±2℃；国标规定单位符号应写在全部数数之后，并在其与数值之间留半角空格，唯一例外是平面角的单位符号°（度）、′（分）、″（秒），与数值之间**不留空**。例如：1527°不写作 1527°；数值与单位之间留半角空格，例：5 m

◇ 与单位相连的数字，最好从小数点向左和向右按三位分成一组，组间留半角空格，但**不得**用逗号、句号等。例如：149600×10^6 m **不应该**写作 149,600×10^{6}m；6894.760Pa **不应该写作 6,894.760Pa

3. SI Prefix

SI 词冠、符号、表示的数值以及中文名称如下

Q	quetta	$10^{3×10}$	昆（它）	M	mega	$10^{3×2}$	兆	n	nano	$10^{-3×3}$	纳（诺）
R	ronna	$10^{3×9}$	容（那）	k	kilo	$10^{3×1}$	千	p	pico	$10^{-3×4}$	皮（可）
Y	yotta	$10^{3×8}$	尧（它）	h	hecto	10^{2}	百	f	femto	$10^{-3×5}$	飞（母托）
Z	zetta	$10^{3×7}$	泽（它）	da	deca	10^{1}	十	a	atto	$10^{-3×6}$	阿（托）
E	exa	$10^{3×6}$	艾（可萨）	d	deci	10^{-1}	分	z	zepto	$10^{-3×7}$	仄（普托）
P	peta	$10^{3×5}$	拍（它）	c	centi	10^{-2}	厘	y	yocto	$10^{-3×8}$	幺（科托）
T	tera	$10^{3×4}$	太（拉）	m	milli	$10^{-3×1}$	毫	r	ronto	$10^{-3×9}$	柔（托）
G	giga	$10^{3×3}$	吉（咖）	μ	micro	$10^{-3×2}$	微	q	quecto	$10^{-3×10}$	亏（科托）

4. 词冠使用规则

◇ SI 词冠优先从 $10^{±3n}$（n 为整数）中选用

◇ 词冠直接与 SI 单位构成新单位：它可以有正负数幂，还可同其他单位构成组合单位

1000m = 1kilometer = 1km (千米)

0.000000000 1s=1nanosecond = 1ns (毫微秒)

1cm^3 = (10^{-2}m)^3 = 10^{-6}m^3

1mm^2/s = (10^{-3}m)^2/s = 10^{-6}m^2/s

◇ 适当选择词冠，最好使数值处在 0.1 - 1000 之间；例如：12000N 可写成 12kN、0.00294m 可写成 2.94mm、1 401Pa 可写成 1.401kPa

◇ 不应使用复合词冠；如：nm（纳米）不用 mμm、Mg（兆克）不用 kkg、mg（毫克）不用 μkg

◇ 词冠应用正体（Roman type）书写，词冠与单位符号之间不留空格；不应随意改变词冠符号以免造成混淆；如：Mg 不得写成 mg，ns 不得写成 Ns

◇ 当词冠号与单位符号相同时，书写时应注意区别

Part B Introduction to Fundamentals of Chemistry and Chemical Engineering

✧ 不允许不与单位组合而单独使用词冠
✧ 除 SI 单位外，还保留一些比较重要的非 SI 单位与 SI 单位并用；主要有：

min (time)	minute（时间）	分
h (time)	hour（时间）	[小]时
d (time)	day（时间）	日
° (plane angle)	degree（平面角）	度
' (plane angle)	minute（平面角）	分
" (plane angle)	second（平面角）	秒
l (volume)	liter（体积）	升
t (mass)	ton（质量）	吨
bar (hydraulic pressure)	bar（液体压力）	巴

✧ SI 词冠可以加在上述单位之前构成新单位；例如：mL (毫升)、mbar (毫巴)等
✧ 上述有些单位也可与 SI 单位及词冠构成新的组合单位；如：t/m^3（吨每三次方米）、km/h（千米每时）

对于科技论文及出版物国际上的趋势是优先使用 SI 单位制。SI 单位有其特定的符号。因此在使用 SI 单位时，应严格遵守使用这些国际符号的规则及建议。

Unit 9
Chemical Engineering

9.1 Overview

Chemical engineering lies at the intersection of science and technology. It's one of the most important engineering disciplines. Take a look at what chemical engineering is, what chemical engineers do, and how to become a chemical engineer. Chemical engineering is applied chemistry. It is the branch of engineering that deals with the design, construction, and operation of machines and plants that perform chemical reactions to solve practical problems or make useful products. It begins in the laboratory, much like science, but progresses through the design and implementation of a full-scale process, its maintenance, and methods for testing and improving it.

Chemical engineering is the profession concerned with the creative application of the scientific principles underlying the transport of mass, energy, and momentum and the physical and chemical transformation of matter. The broad implications of this definition have been justified over the past several decades by the types of problems that chemical engineers have solved, although the profession has focused its attention primarily on the chemical process industries[1]. As a result, chemical engineers have been more traditionally defined as those applied scientists trained to deal with the research, development, design, and operating problems of the chemical, petroleum, and related industries. Experience has shown that the principles required to meet the needs of the process industries are applicable to a much broader class of problems, and the modern chemical engineer brings his established tools to bear on such new areas as the environmental and life sciences.

Chemical engineering developed as a distinct discipline in the twentieth century in response to the needs of a chemical industry that could no longer operate efficiently with manufacturing processes that were, in many cases, simply larger versions of laboratory equipment[2]. Thus, the initial focus of the profession was on the general issue of how to use the results of laboratory experiments to design process equipment capable of meeting industrial production rates. This naturally led to the characterization of design procedures in terms of unit operations, those elements common to many different processes. The basic unit operations include fluid flow, heat exchange, distillation, extraction, etc. A typical manufacturing process will consist of combinations of unit operations. Therefore, the ability to design each of the unit operations on a production scale would provide the means to design the entire process.

The unit operations concept dominated chemical engineering education and practice until the mid-1950s, when a movement away from this equipment-oriented philosophy toward an engineering science approach began. This approach holds that the unifying concept is not specific processing operations, but rather an understanding of the fundamental phenomena of mass, energy, and momentum transport that are common to all unit operations, and it is argued that a focus on unit operations obscures the similarity of many operations at a fundamental level[3].

Although there is no real conflict between the goals of the unit operations and engineering science approaches, the latter has tended to emphasize mathematical skills and deemphasize the design aspects of engineering education. Such a conflict need not exist, and recent educational efforts have been directed toward developing the skills that will enable creative engineering use of the fundamentals, or a synthesis of the engineering science and unit operations approaches. An essential skill in achieving this goal is the ability to express engineering problems meaningfully in precise quantitative terms. Only in this way can the chemical engineer correctly formulate, interpret, and apply fundamental experiments and physical principles in real-world applications outside the laboratory. This skill, which is distinct from mathematical skill, is what we call analysis.

In short, chemical engineering is an engineering discipline that emphasizes chemistry. Those involved in chemical engineering should have a working knowledge of chemistry and chemical processes. Chemical engineering also involves reaction engineering. Chemical engineering involves mass and heat transfer of chemical products. Chemical engineering is also the field of fluid flow. Those who work in the field of chemical engineering are involved in a variety of areas. Chemical engineering is found in research and development, process design, plant design and construction, sales, management.

Chemical engineering has been responsible for products and processes that modern society enjoys today. Chemical engineering has ranged from petroleum and petrochemical products to products derived from natural minerals and deposits. Chemical engineering has produced gasoline, packaging film, and is working on developing fuel cells for the future electric car.

This unit summarizes only the concepts related to the unit operations of a chemical engineering process and the basic chemistry knowledge involved in chemical engineering.

9.2 The Difference between Chemistry and Chemical Engineering

Let's start by looking at the differences between science and engineering. The difference between engineering and science centers on methodology, focusing on the following categories: (1) Purpose, where scientists always want to understand how the universe works, while engineers just want to design a mechanism or system that works according to known laws and applied to specific needs, and (2) Routine, where scientists always develop a theory and test it, while engineers instead just want to gather information, make a plan, build and test a prototype, iterate out the kinks, and go to production and marketing.

Although there is some overlap between chemistry and chemical engineering, the courses you take, the degrees you earn, and the jobs you do are quite different. Let's take a look at what chemists and chemical engineers study and what they do. The big difference between chemistry and chemical engineering has to do with originality and scope.

Chemists are more likely to develop new materials and processes, while chemical engineers are more likely to take those materials and processes and make them larger or more efficient. Chemists take courses in all major branches of chemistry, general physics, mathematics up to calculus and possibly differential equations, and may take courses in computer science or programming. Chemists also typically take "core" courses in the humanities. Chemists with a bachelor's degree typically work in laboratories. They may contribute to research and development or perform sample analysis. Master's-level chemists do the same type of work, plus they may supervise research. Chemists with a Ph.D. manage and conduct research, or they may teach chemistry at the college or graduate level. Most chemists pursue advanced degrees and may intern with a company before joining a company. It's much harder to get a good chemistry job with a bachelor's degree than with the specialized training and experience gained during graduate school.

Most chemical engineers have a bachelor's degree in chemical engineering. Master's degrees are also popular, while Ph.D.s are rare compared to chemistry degrees. Chemical engineers take an exam to become licensed professional engineers. After gaining enough experience, they may become professional engineers (P.E.). Chemical engineers take most of the chemistry courses taken by chemists, plus engineering courses and additional math. Additional math courses include differential equations, linear algebra, and statistics. Common engineering courses include fluid dynamics, mass transfer, reactor design, thermodynamics, and process design. Engineers may take fewer core courses, but often take ethics, economics, and business classes. Chemical engineers work on research and development teams, process engineering in a plant, project engineering, or management. Similar jobs are available at the entry-level and graduate levels, although engineers with master's degrees often find themselves in management. Many chemical engineers have started their own businesses.

9.3 Future Opportunities in Chemical Engineering

In many ways, we have moved beyond the traditional definition of the chemical industry. Changes in the global business environment are affecting all industries, and I would like to look at what this means for chemical companies and the people who work for them — especially scientists and engineers.

There is a growing recognition of the broad application of chemistry and chemical engineering. Chemistry, in interaction with other disciplines, provides the fundamental knowledge needed to address many of society's needs. These include new materials for the aerospace, automotive, and electronics industries; fundamental data for designing effective environmental controls; and understanding life processes important to agriculture and health care. The maturation of the commodity chemicals, plastics, and apparel fibers businesses, increased international competition, and issues of environmental quality and toxic materials have led many to take a defensive view of our industry. If we are not careful, such a view could stifle creativity. In fact, the opportunities for new businesses based on the chemical sciences are great. But to take advantage of these opportunities, we must be willing to move out of traditional areas and into new ones.

As the demand for finite resources increases and the world's population grows rapidly, chemical engineering will be at the forefront of solving the world's greatest challenges. Chemical engineering plays a key role in areas such as health care, energy development, waste management, food processing, and national security.

Chemical engineering draws on many different fields, not only the chemical sciences, but also mathematics, physics, materials science, biology, and data science. Chemical engineering applies the molecular sciences to create the essentials of life. Over the past three decades, chemical engineering has undergone tremendous changes and advances, and these changes have and will continue to affect how we view research priorities, education, and the practice of chemical engineering. There is a growing recognition of the broad opportunities for the application of chemical engineering.

9.4 Introduction to Basic Chemical Engineering Concepts

9.4.1 Heat Transfer and Heat Exchangers

Heat transfer is the movement of heat due to a temperature difference between a system and its surroundings. Heat is always transferred in one way or another wherever there is a temperature difference. Just as water always flows downward to the lowest possible level, heat, if left to its own devices, is transferred from hot objects to cold objects, always warming the cold objects at the expense of the warmer ones. The rate at which heat flows depends on the size of the temperature difference as well as the properties of the material through which it must flow.

There are three ways in which heat is transferred. Since heat itself is the energy of molecular activity, the simplest mode of heat transfer, called conduction, is the direct communication of molecular disturbance through a substance by the collision of neighboring molecules. Metals contain what are called "free" electrons, which make them good conductors of electricity; these electrons also contribute to the conduction of heat, so metals have high thermal conductivities.

Convection is the transfer of heat from one place to another by the actual movement of the hot material. Heat is also transferred by a combination of radiation and absorption. Radiation is the conversion of thermal energy into radiant energy, similar in nature to light. While in the form of radiation, the energy can travel an enormous distance before being absorbed or converted back to heat. For example, energy radiated from the surface of the sun is converted to heat at the surface of the earth only eight minutes later.

In a chemical process, it is often necessary to change the temperature of a fluid stream. This can be done by passing the fluid through a **heat exchanger** where it is in thermal contact (but usually not in direct physical contact) with another fluid at a different temperature. The most common arrangement used to achieve this is to pass one of the fluids through a metal tube immersed in the other fluid. The fluid to be heated passes through the inner tube, and the hot fluid providing the heat passes through a coaxial outer tube.

Mounting. Because of the mechanical difficulties of concentrically supporting one tube within another, a more common physical arrangement is known as a shell-and-tube heat exchanger, in which one fluid passes in parallel through a large number (often up to 100) of tubes in parallel, supported on a large cylindrical vessel. The other fluid passes through the cylinder, called the shell side, and is forced to flow back and forth across the tubes in a zigzag pattern. This creates a great deal of turbulence, which increases the rate of heat transfer and also ensures good mixing. The temperature of the fluid on the shell side is therefore nearly constant at each section, so the only variation is in the axial direction. Most heat exchangers are liquid-liquid, but gases and non-condensing vapors can be handled.

9.4.2 The Material Balance and the Energy Balance

If matter may neither be created nor destroyed, the total mass of all materials entering an operation is equal to the total mass of all materials leaving the operation, except for any material that may be retained or accumulated in the operation. This principle is used to calculate the yield of a chemical reaction or an engineering operation.

In continuous operations, material is usually not accumulated in the operation, and a material balance simply consists of charging (or debiting) the operation with all material entering it and crediting the operation with all material leaving it, just as any accountant would do. The result must be a balance.

As long as the reaction is chemical and does not destroy or create atoms, it is correct and often very convenient to use atoms as the basis for the material balance. The material balance may be made for the whole plant or for any part of it as a unit, depending on the problem at hand. Similarly, an energy balance can be made around any plant or unit operation to determine the energy required to continue the operation or maintain desired operating conditions. The principle is just as important as the material balance and is used in the same way. The important point to remember is that all energy of all types must be included, even though it may be converted to a single equivalent.

9.4.3 The Ideal Contact and the Rates of an Operation

Whenever the materials to be processed are in contact for any length of time under specified conditions, such as temperature, pressure, chemical composition, or electrical potential, they tend to approach a certain equilibrium state determined by the specified conditions. In many cases, the rate of approach to these equilibrium conditions is so rapid, or the length of time is so great, that the conditions are practically reached at each contact. Such a contact is called an equilibrium or ideal contact. Calculating the number of ideal contacts is an important step in understanding those unit operations that involve the transfer of material from one phase to another, such as leaching, extraction, absorption, and dissolution.

In most operations, equilibrium is not reached either because there is not enough time or because it is not desired. Once equilibrium is reached, no further change can occur and the process stops, but the engineer must continue the process. For this reason, rate operations, such as the rate of energy transfer, the rate of mass transfer, and the rate of chemical reaction, are of the greatest importance and interest. In all of these cases, the rate and direction depend on a difference in potential or driving force. The rate can usually be expressed as proportional to a drop in potential divided by a resistance. An application of this principle to electrical energy is the familiar Ohm's law for steady or direct current.

Part B Introduction to Fundamentals of Chemistry and Chemical Engineering   151

In solving rate problems, such as heat or mass transfer, with this simple concept, the major difficulty is the evaluation of the resistance terms, which are usually calculated from an empirical correlation of many determinations of transfer rates under different conditions[4].

The basic concept that the rate depends directly on a potential drop and inversely on a resistance can be applied to any rate operation, although the rate may be expressed in different ways, special coefficients for special cases.

9.4.4 Chemical Engineering Process Related to Solids

(1) Size reduction and the application of size reduction

Size reduction is the term used to describe all the ways in which particles of solids are cut or broken into smaller pieces. Throughout the process industries, solids are reduced by different methods for different purposes. Lumps of raw ore are crushed to workable size, synthetic chemicals are ground into powders, and sheets of plastic are cut into tiny cubes or diamonds. Commercial products must often meet strict specifications for the size and sometimes the shape of the particles they contain. Reducing particle size also increases the reactivity of solids, allows unwanted constituents to be separated by mechanical methods, reduces the bulk of fibrous materials for easier handling; and so on.

Solids can be broken in eight or nine different ways, but only four of them are commonly used in size reduction equipment. They are (1) compression, (2) impact, (3) abrasion, and (4) cutting. A nutcracker, a sledgehammer, a file, and a pair of scissors are examples of these four types of action. In general, compression is used for coarse reduction of hard solids to produce relatively few fines: impact produces coarse, medium, or fine products; attrition produces very fine products from soft, nonabrasive materials. Cutting yields a precise particle size and occasionally a distinct shape, with minimal or no fine particles.

Comminution is a general term for size reduction. Crushers and grinders are types of size reduction equipment. An ideal crusher or mill would (1) have a large capacity, (2) require a small amount of power per unit of product, and (3) produce a product of the desired single size or size distribution. The usual method of studying the performance of process equipment is to set up a business operation as a standard, compare the characteristics of the actual equipment with those of the ideal unit, and explain the difference between the two. When this method is applied to crushing equipment, the discrepancies between the ideal and the actual are considerable, and the gaps have not been fully explained, even theoretically. On the other hand, useful quantitative information can be obtained from the incomplete theory now available.

The capacities of shredders are best discussed when the individual types of equipment are described. However, the fundamentals of product size and shape and energy requirements are common to most machines and can be discussed more generally.

In the chemical industry, size reduction is typically performed to increase surface area. In fact, in most reactions involving solid particles, the rate is directly proportional to the area of contact with a second phase. Thus, the rate of combustion of solid particles is proportional to the area presented to the gas, although a number of secondary factors may also be involved, such as the free flow of the gas being impeded by the higher flow resistance of a bed of small particles. Again, in leaching, not only is the extraction rate increased due to the increased contact area between the solvent and the solid, but the distance that the solvent has to penetrate into the particles to access the more remote solute pockets is also reduced. This factor is also important in the drying of porous solids, where the reduction in size causes both an increase in area and a reduction in the distance the moisture must travel within the particles to reach the surface; in this case, the capillary forces acting on the moisture are also affected.

There are a number of other reasons for size reduction. For example, it may be necessary to break a material into very small particles in order to separate two components, especially if one is dispersed in small isolated pockets. In addition, the properties of a material can be significantly affected by particle size; for example, the chemical reactivity of fine particles is greater than that of coarse particles, and the color and hiding power of a pigment is significantly affected by particle size[5]. Again, much more intimate mixing of solids can be achieved when the particle size is small.

(2) Characterization of solid particles and screening

Individual solid particles are characterized by size, shape and density. Particles of homogeneous solids have the same density as the bulk material. Particles obtained by breaking up a composite solid, such as a metal-bearing ore, have different densities, usually different from the density of the bulk material. Size and shape are easily specified for regular particles, such as spheres and cubes, but for irregular particles (such as sand grains or mica flakes), the terms "size" and "shape" are not clear.

In general, "diameter" can be specified for all equidimensional particles. Particles that are not equidimensional, that is, longer in one direction than in another, are often characterized by the second longest major dimension. For example, for needle-like particles, D_p would refer to the thickness of the particles, not their length.

By convention, particle sizes are expressed in different units depending on the size range. Coarse particles are measured in inches or centimeters; fine particles are measured in mesh size; very fine particles are measured in microns or nanometers. Ultrafine particles are sometimes described in terms of surface area per unit mass, usually in square meters per gram.

Standard test sieves are used to measure the size (and size distribution) of particles in the range of approximately 0.0015 to 3 inches (1 inch=0.0254m). Test sieves are made of woven wire cloth with carefully standardized meshes and dimensions. The openings are square. Each screen is specified in meshes (100 meshes = 150 μm)per inch. However, the actual openings are smaller than those corresponding to the mesh numbers because of the thickness of the wires. The area of the apertures in any screen in the standard series is exactly twice the area of the apertures in the next smaller screen. Therefore, the ratio of the actual mesh size of any screen to that of the next smaller screen is $\sqrt{2}$ = 1.414. Intermediate screens are available for tighter sizing, each of which has a mesh size $\sqrt[4]{2}$, or 1.189 times that of the next smaller standard screen. These intermediate screens are not normally used.

Screening is a method of separating particles by size only. In industrial screening, solids are dropped or thrown onto a screen surface. The undersize, or fines, pass through the screen apertures; the oversize, or tails, do not. A single screen can make only one separation into two fractions. These are called unsized fractions because although either the upper or lower limit of the particle sizes they contain is known, the other limit is unknown. Material passed through a series of screens of different sizes is separated into size fractions, i.e. fractions in which both the maximum and

Part B Introduction to Fundamentals of Chemistry and Chemical Engineering   153

minimum particle sizes are known. Screening is occasionally wet, but much more commonly dry. The discussion in this section is limited to the screening of dry particulate solids.

Industrial screens are made of metal bars perforated on slotted metal plates, woven wire cloth, or woven fabric such as silk. Metals used include steel, stainless steel, bronze, copper, nickel, and monel. The mesh size of woven screens ranges from 4 to 400 mesh, but screens finer than 100 to 150 mesh are rarely used. For very fine particles, other methods of separation are usually more economical. Many varieties and types of screens are available for different purposes.

In most screens, the particles fall through the screen opening by gravity. In a few designs not discussed here, they are pushed through by a brush or by centrifugal force. Coarse particles fall quickly and easily through large openings in a stationary surface; for finer particles, the screen surface must be agitated in some way. Common methods are to rotate a cylindrical screen about a horizontal axis; or, in the case of flat screens, to shake, gyrate, or vibrate them mechanically or electrically.

9.4.5 Fluidization

A fluidized bed is a relatively stable mixture of a fluid, usually a gas, and finely divided solids that is intermediate between a fixed bed and pneumatic transport conditions. Its behavior is similar to that of a boiling liquid. Fluid velocities are in the range that will lift and agitate the appropriately sized particles without completely carrying the particle out of the containment vessel. Solid particle sizes in the range of 3/8 inch to 10 microns (μm) can be handled satisfactorily, with some size gradations preferred for smooth fluidization.

Fluidized bed techniques to create intimate contact between small solid particles and fluids are standard design practice where these characteristics are desired as follows.

◆ High rates of reaction between solids and gases under controlled conditions: exposed reaction surface area is inversely proportional to particle size, so a large reaction surface area per unit reactor volume is required; violent agitation that occurs between the carrier gas and the solids produces high rates of mass and heat transfer under very uniform temperature conditions

◆ Uniform reactor temperatures: fluidized beds have high thermal conductivities compared to the carrier gas alone, resulting in very uniform bed temperature profiles compared to fixed bed operation

◆ High heat transfer rates: heat transfer rates from the fluid-solid mixtures to any transfer surface within the moving zone are quite high as compared to gases alone, with values as high as 200 Btu · h^{-1} · ft^{-2} · $°F^{-1}$ [1 Btu= 1055.06J, 1 ft = 0.3048m, $t/°C=\frac{5}{9}(t/°F-32)$]being reported; thus, minimal heat transfer surface is required when using fluidized bed reactors

◆ Ease of solids transport: the fluidized solids behave as a liquid, so they can be transported from one reactor to another to add or remove heat, or to perform another step in the process

When a fluid is passed upward through a bed of fine particles, at a low flow rate, the fluid will simply percolate through the voids between the stationary particles. This is a fixed bed. As the flow rate increases, the particles move apart and some of them are seen to vibrate and move in confined regions. This is the expanded bed.

At an even higher velocities, a point is reached where the particles are all just suspended in the upward flowing gas or liquid. At this point, the fractional force between the particle and the fluid is equal to the weight of the particle, the vertical component of the compressive force between adjacent particles disappears, and the pressure drop across any section of the bed is approximately equal to the weight of the fluid and particles in that section. The bed is considered to be just fluidized and is referred to as an incipiently fluidized bed or a bed at minimum fluidization.

In liquid-solid systems, an increase in flow rate beyond minimum fluidization usually results in a smooth, progressive expansion of the bed. Gross flow instabilities are damped and remain small, and large-scale bubbling or heterogeneity are not observed under normal conditions. Such a bed is called a highly fluidized bed, a homogeneously fluidized bed, a smoothly fluidized bed, or simply a fluidized bed.

Gas-solid systems generally behave quite differently. As the flow rate increases beyond minimum fluidization, large instabilities with bubbling and channeling of the gas are observed. At higher flow rates, the agitation becomes more violent and the movement of the solids becomes more vigorous. In addition, the bed does not expand much beyond its volume at minimum fluidization. Such a bed is called an aggregative fluidized bed, a heterogeneously fluidized bed, a bubbling fluidized bed, or simply a gas fluidized bed.

In a few rare cases, liquid-solid systems do not fluidize smoothly and gas-solid systems do not bubble. At present, such beds are only laboratory curiosities of theoretical interest.

Both gas and liquid fluidized beds are considered to be dense phase fluidized beds as long as there is a fairly well defined upper boundary or surface to the bed. However, at a sufficiently high fluid flow rate, the terminal velocity of the solids is exceeded, the upper surface of the bed disappears, entrainment becomes significant, and solids are carried out of the bed with the fluid flow. In this state we have a disperse, dilute or lean phase fluidized bed with pneumatic transport of solids.

Compared to other methods of gas-solid contacting, gas-fluidized beds have some rather unusual and useful properties. These are not shared to the same extent by liquid-solid systems, and thus virtually all important industrial applications of fluidization are gas-fluidized systems.

9.4.6 Supercritical Fluids and Supercritical Fluid Extraction

Supercritical fluids (SCFs) are increasingly replacing organic solvents in industrial purification and recrystallization operations due to regulatory and environmental pressures on hydrocarbon and ozone-depleting emissions. Figure 9.1 shows a typical phase diagram of SCF. SCF-based processes have helped eliminate the use of hexane and methylene chloride as solvents. With increased scrutiny of solvent residues in pharmaceuticals, medical devices and netraceuticals, and stricter regulations on VOC and ODC emissions, the use of SCFs is rapidly expanding in all industrial sectors.

Supercritical fluid extraction plants operate at throughputs of 100 000 000 lb/a (1lb = 0.4536 kg) or more in the food industry. Coffee and tea are decaffeinated by supercritical fluid extraction, and most major breweries in the world use flavors extracted from hops by supercritical

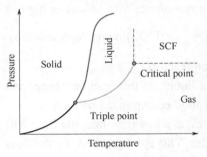

Figure 9.1　Phase diagram of SCF

fluid extraction.

SCF processes are being commercialized in the polymer, pharmaceutical, and specialty lubricants and fine chemicals industries. SCFs are advantageously used to improve product performance to levels that cannot be achieved by traditional processing technologies, and such applications of SCFs offer the potential for both technical and economic success. The critical point (CP) marks the end of the vapor-liquid coexistence curve. A fluid is said to be supercritical when the temperature and pressure are higher than the corresponding critical values. Above the critical temperature, there is no phase transition because the fluid cannot undergo a transition to a liquid phase regardless of the pressure applied.

A supercritical fluid (SCF) is characterized by physical and thermal properties between those of a pure liquid and a gas. The density of the fluid is a strong function of temperature and pressure. The diffusivity of SCF is much higher than that of a liquid, and SCF readily penetrates porous and fibrous solids. As a result, SCF can provide good catalytic activity. There are drastic changes in some important properties of a pure liquid as its temperature and pressure are increased near the thermodynamic critical point. For example, under conditions of thermodynamic equilibrium, the visual distinction between the liquid and gas phases, as well as the difference between the liquid and gas densities, disappears at and above the critical point. Similar drastic changes occur in the properties of a liquid mixture as it approaches the thermodynamically critical region of the mixture.

Other properties of a liquid fuel that change greatly near the critical region are thermal conductivity, surface tension, heat capacity at constant pressure, and viscosity. When comparing a liquid sample with a supercritical fluid (SCF) sample of the same fuel, both of the same density, the thermal conductivity and diffusivity of the SCF are higher than the liquid, its viscosity is much lower, while its surface tension and heat of vaporization have completely disappeared. These drastic changes make a supercritical fuel significantly preferable to a liquid fuel of the same density. Moreover, the combustion phenomena resulting from a supercritical fuel are expected to be quite different from those of a liquid fuel.

SCF applications include oil shale organic recovery, biological fluid separation, bioseparation, petroleum recovery, crude asphalting and dewaxing, coal processing (reactive extraction and liquefaction), selective extraction of fragrance, oils and contaminants from agricultural and food products, pollution control, combustion, and many other applications.

Supercritical Fluid Extraction (SFE) is based on the fact that near the critical point of the solvent, its properties change rapidly with only small changes in pressure. Supercritical fluids can be used to extract analytes from samples. The main advantages of using supercritical fluids for extractions are that they are inexpensive, extract analytes faster, and are more environmentally friendly than organic solvents. For these reasons, supercritical fluid CO_2 is the most commonly used reagent as a supercritical solvent.

The advantages of SFE include the following six aspects:

✧ SCFs have solvating power similar to liquid organic solvents, but with higher diffusivity, lower viscosity, and lower surface tension

✧ Because the solvating power can be adjusted by changing pressure or temperature, the separation of analytes from the solvent is fast and easy

✧ By adding modifiers to an SCF (such as methanol to CO_2), its polarity can be changed for more selective separation performance

◇ Industrial processes involving food or pharmaceuticals do not have to worry about solvent residues as they would with a "typical" organic solvent

◇ Candidate SCFs are generally inexpensive, simple, and many are safe. Disposal costs are much lower, and the fluids can be easily recycled in industrial processes

◇ SCF technology requires sensitive process control, which is challenging. In addition, the phase transitions of the solute/solvent mixture must be measured or predicted with high accuracy. In general, the phase transitions in the critical region are quite complex and difficult to measure and predict. Our research has provided much insight into these phenomena

9.4.7 Extraction and Liquid-Liquid Extraction

Solvent extraction is the transfer of a solute species from its original location to a solvent, known as the extraction solvent. The pure solute may be a liquid or a solid at the operating temperature, but this is unimportant because the solute is initially either in solution or associated with a solid. If the solute is in solution, the extraction process is called liquid-liquid extraction, and the extracting solvent must be substantially immiscible with the original solvent. On the other hand, if the solute is part of a solid (which need not be "dry"), the process is called solid-liquid extraction and will be discussed later.

In liquid-liquid extraction, the extracting solvent must have a sufficiently selective affinity for the appropriate solute, sometimes in the company of materials other than the original solvent. This selectivity is very important because the essence of liquid-liquid extraction is the separation of a particular solute from other materials by selective transfer to the extracting solvent. It must be remembered that the extracted solute is not always the most valuable product of the separation process; the aim may be to purify the original solvent by removing an unwanted solute, or perhaps to remove one of two solutes from the original solution.

Separation by solvent extraction is often attractive in circumstances where distillation is inappropriate. For example, if the solute is heat sensitive or present in very low concentrations, liquid-liquid extraction may be appropriate.

In solid-liquid extraction, a component of a solid is transferred to an extracting solvent and separated from the rest of the solid. The extracted material is not necessarily a solid, but may be present in the bulk solid in liquid form. Some solids even have an intrinsic solvent content that becomes apparent during the separation process. For example, sugar beets, are extracted with water and are themselves approximately 30% water by mass.

Because a solid does not flow like the liquids in liquid-liquid extraction, the equipment for solid-liquid extraction is different from that for liquid-liquid extraction.

Liquid-liquid extraction is used when distillation and rectification are difficult or ineffective. Low boiling mixtures or substances that cannot withstand the vaporization temperature even under vacuum can often be separated by extraction. Extraction exploits differences in the solubility of the components rather than differences in their volatility. Since solubility depends on chemical properties, extraction uses chemical differences rather than vapor pressure differences. When either distillation or extraction can be used, the choice is usually distillation, although heat and cooling are required. If extraction is used, the solvent must be recovered for reuse (usually by distillation), and the combined operation is more complicated and often more expensive than simple distillation without extraction. For many problems, the choice between methods should be based on a comparative study of both extraction and distillation.

Extraction can be used to separate more than two components. Some applications require mixtures of solvents rather than a single solvent. These more complicated methods are not discussed here. In liquid-liquid extraction, as in gas absorption and distillation, two phases must be brought into good contact to allow transfer of material and then separated. In absorption and distillation, the mixing and separation is simple and fast. In extraction, however, the two phases have comparable densities, so the energy available for mixing and separation when using gravity flow is small, much smaller than when one phase is a liquid and the other is a gas. The two phases are often difficult to mix and difficult to separate. The viscosities of the two phases are also relatively high, and the linear velocities through most extraction equipment are low. Therefore, in some types of extractors, the energy for mixing and separation is supplied mechanically. Extractors can be operated in batch or continuous mode. Most extraction equipment is continuous.

9.4.8 Evaporation and Crystallization

The goal of evaporation is to concentrate a solution consisting of a nonvolatile solute and a volatile solvent. In the vast majority of evaporations, the solvent is water. Evaporation is performed by evaporating a portion of the solvent to produce a concentrated solution or thick liquor. Evaporation differs from drying in that the residue is a liquid (sometimes a highly viscous one) rather than a solid; it differs from distillation in that the vapor is a mixture, no attempt is made in the evaporation step to separate the vapor into fractions; it differs from crystallization in that the emphasis is on concentrating a solution rather than forming and building crystals. In certain situations, such as the evaporation of brine to produce common salt, the line between evaporation and crystallization is far from sharp. Evaporation sometimes produces a slurry of crystals in a saturated mother liquor.

Usually in evaporation, the thick liquor is the valuable product and the vapor is condensed and discarded. Mineral-bearing water is often evaporated to produce a solid-free product for boiler feed, special process requirements, or human consumption. This technique is often referred to as water distillation, but technically it is called evaporation. In one particular situation, however, the reverse is true. A large-scale evaporation process is being developed and used to produce potable water from seawater. Here, the condensed water is the desired product. Only a fraction of the total water in the feed is recovered and the rest is discarded.

Crystallization from liquid solution is important industrially because of the variety of materials that are marketed in the crystalline form. Its wide use is based on the fact that a crystal formed from an impure solution is itself pure (unless mixed crystals occur) and that crystallization provides a practical method of obtaining pure chemical substances in a satisfactory condition for packaging and storage.

Clearly, good yield and high purity are important objectives in the operation of a crystallization process, but these are not the only factors to consider. The appearance and size range of a crystalline product are also important. In particular, it is necessary for the crystals to be of a reasonable and uniform size. If they are to be further processed, reasonable size and uniformity is desirable for washing, filtering, reacting with other chemicals, transporting and storing the crystals. If the crystals are to be marketed as a finished product, customers require the individual crystals to be strong, non-aggregated, uniform in size, and not clumped in the package. For these reasons, crystal size distribution (CSD) must be controlled and is a primary objective in the design and operation of crystallizers.

In general, crystallization can be analyzed in terms of purity, yield, energy requirements, and rates of crystal formation and growth.

9.4.9 Dust Removal and Centrifugal Settling Process

Generally, gas must be cleaned of dust for one or more of the following three purposes:

◇ The dust contained in the gas may be valuable and separated as another product of the process.

◇ The gas itself may be required for use in another process.

◇ It may be an effluent gas that needs to be cleaned before being released into the atmosphere to prevent nuisance or damage to amenities.

Gas cleaning can be a very expensive process, especially where almost complete removal of all dust is required, and great care must be taken to select the most economical equipment for a particular purpose, taking into account capital and operating costs and the value of the particles collected. It is therefore necessary to have a good knowledge of the main features and applications of the various types of gas cleaning equipment available.

The simplest method of removing particles from a moving gas stream is to allow them to settle under the force of gravity. Large particles will often do this in a simple settling chamber. It is considered to be an efficient collector of coarse particles. Cyclones are another type of removal system where particles are removed from spinning gases by centrifugal forces. They are simple to construct and have no moving parts. The centrifugal force on particles in a spinning gas stream is much greater than gravity, therefore cyclones are more effective at removing much smaller particles than gravitational settling chambers and require much less space to handle the same volumes of gas. On the other hand, the pressure drop in a cyclone is greater and the power consumption is much higher.

A given particle in a given fluid will settle under gravity at a fixed maximum rate. To increase the settling rate, the gravity force acting on the particle can be replaced by a much stronger centrifugal force. Centrifugal separators have largely replaced gravity separators in production operations because of their greater effectiveness with fine droplets and particles and their much smaller size for a given capacity. Most centrifugal separators used to remove particles from gas streams have no moving parts. The cyclone is a typical example. It consists of a vertical cylinder with a conical bottom, a tangential inlet near the top, and a dust outlet at the bottom of the cone. The inlet is usually rectangular. The outlet tube is extended into the cylinder to prevent short circuiting of the gas from inlet to outlet.

The incoming dust-laden gas is given a rotational motion as it enters the cylinder. The resulting vortex develops a centrifugal force that causes the particles to be thrown radically against the wall. Basically, a cyclone is a settling device that uses a strong centrifugal force acting radially instead of a relatively weak gravitational force acting vertically. The centrifugal force in a cyclone ranges from 5 times the force of gravity in large, low-velocity units to 2500 times the force of gravity in small, high-pressure units.

The path of the gas in the cyclone follows a downward vortex, or spiral, adjacent to the wall and reaches the bottom of the cone. The gas stream then moves upward in a tighter spiral, concentric with the first, and exits through the outlet pipe, still swirling. Both spirals rotate in the same direction. Inside the cyclone, dust particles are deposited against the wall, along which they slide down to the bottom of the cone. A multicyclone is an assembly of small tubular cyclones. It is often used to treat a large amount of gas with small particles.

9.4.10 Batch and Continuous Process

Many industrial operations can be performed in one of two ways, which can be referred to as batch and continuous operations, e. g. the simple case of heating a solution with steam. In the batch process, the solution is pumped into a tank at its original temperature, and then heated as a batch by admission of steam to a surrounding jacket or to internal coils. When the solution in the tank reaches the desired temperature, it is discharged and a new batch of cool solution is added. In the continuous process, the solution can be passed slowly but continuously through a pipe coil heated by steam, with the flow rate adjusted so that the solution leaves the outlet end of the coil at the desired temperature.

Although the continuous process requires more carefully designed equipment than the batch process, it usually requires less space, is more easily integrated with other continuous steps, and can be operated at any prevailing pressure with venting to atmospheric pressure[6]. The temperature of each part of the equipment remains essentially constant during operation, thus avoiding the fluctuating, which is unavoidable in the batch process. Continuous operation of processes has many advantages and is ordinarily a goal in engineering design. Batch operation is often found in experimental and pilot plant operations.

9.4.11 Chemical Manufacturing Process

The basic components of a typical chemical process of producing a product from the raw materials are shown in Figure 9.2, which represents a generalized process, not all stages will be required for a particular process, and the complexity of each stage will depend on the nature of the process. Chemical engineering design is concerned with the selection and arrangement of the stages and the selection, specification, and design of the equipment required to perform the functions of the stages.

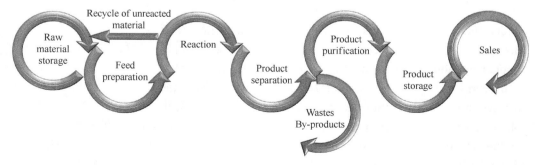

Figure 9.2   General chemical manufacturing process diagram

✧ Raw material storage. Unless the raw materials (also called essential materials or feedstocks) are supplied as intermediates from a neighboring plant, some provision must be made for several days or weeks of storage to smooth out fluctuations and interruptions in supply. Even if the materials come from an adjacent plant, some provision is usually made to hold a few hours or even days' supply to decouple the processes. The amount of storage required depends on the nature of the raw materials, the method of delivery, and the assurance of continuity of supply. If materials are delivered by ship (tanker or bulk carrier), several weeks of storage may be required, whereas if they are received by road or rail in smaller batches, less storage is required.

✧ Feed preparation. Some purification and preparation of the raw materials are usually required before they are sufficiently pure or in the proper form to be fed to the reaction stage. For example, acetylene produced by the carbide process contains arsenic and sulfur compounds and other impurities that must be removed by washing with concentrated sulfuric acid (or other processes) before it is sufficiently pure to react with hydrochloric acid to produce dichloroethane. Liquid feeds must be evaporated before being fed to gas-phase reactors, and solids may require crushing, grinding, and screening.

✧ Reactor. The reaction stage is the heart of the chemical manufacturing process. In the reactor, raw materials are brought together under conditions that promote the production of the desired product; inevitably, by-products and unwanted compounds (impurities) are also formed.

✧ Product separation. In this first stage after the reactor, the products and by-products are separated from any unreacted material. If sufficient, the unreacted material is returned to the reactor. They can be returned directly to the reactor or to the feed cleaning and preparation stage. The by-products may also be separated from the products at this stage.

✧ Purification. Prior to sale, the main product will usually require purification to meet the product specifications. If produced in economic quantities, the by-products may also be purified for sale.

✧ Product storage. Some inventory of finished product must be maintained to match production with sales. Provisions for product packaging and transportation are also required, depending on the nature of the product. Liquids are normally dispatched in drums and bulk tankers (road, rail and sea), while solids in sacks, cartons or bales. Stock levels depend on the nature of the product and the market.

New Words and Expressions

intersection	[ˌɪntəˈsekʃn]	n. 交接，相交
discipline	[ˈdɪsəplɪn]	n. 训练，科目，学科
implementation	[ˌɪmplɪmenˈteɪʃ(ə)n]	n. 实施，执行
petroleum	[pəˈtrəʊliəm]	n. 石油，原油
momentum	[məˈmentəm]	n. 冲力，势头，动量，冲量
phenomena	[fəˈnɒmɪnə]	n. 现象（phenomenon 的复数形式）
mathematical	[ˌmæθəˈmætɪk(ə)l]	adj. 有关数学的，精确的，完整的
deemphasize	[diːˈemfəsaɪz]	v. 使不重要，不再给予强调
petrochemical	[ˌpetrəʊˈkemɪkl]	n. 石油化学产品；adj. 石油化工的

英文	音标	释义
prototype	[ˈprəutətaɪp]	n. 原型，雏形，典型，范例；v. 制作（产品的）原型
linear algebra		n. 线性代数
thermodynamics	[ˌθɜːməudaɪˈnæmɪks]	n. 热力学
disturbance	[dɪˈstɜːbəns]	n. 湍动，扰动
convection	[kənˈvekʃn]	n. 对流，传送
tremendous	[trəˈmendəs]	adj. 巨大的，极好的，精彩的
radiation	[ˌreɪdiˈeɪʃn]	n. 辐射，放射线
fluid	[ˈfluːɪd]	n. 流体；adj. 流体的（fluidize v. 使液化；fluidization n. 流体化）
coaxial	[kəuˈæksɪəl]	adj. 同轴的，共轴的
mechanical	[məˈkænɪkl]	adj. 机动的，机械的
turbulence	[ˈtɜːbjələns]	n. 骚乱，动荡，湍流，紊流
zigzag	[ˈzɪɡzæɡ]	n. 之字形，锯齿形线条，急转弯
accumulated	[əˈkjuːmjəleɪtɪd]	adj. 累计的
equilibrium	[ˌiːkwɪˈlɪbriəm]	n. 平衡，均衡，心理平衡
specification	[ˌspesɪfɪˈkeɪʃn]	n. 详述，规格，说明书，规范
reduction	[rɪˈdʌkʃ(ə)n]	n. 减小，降低，减价，缩图，缩版，归纳，简化
compression	[kəmˈpreʃ(ə)n]	n. 压紧，压缩
abrasion	[əˈbreɪʒ(ə)n]	n. 磨损，磨耗，擦伤
nutcracker	[ˈnʌtkrækə(r)]	n. 胡桃钳，星鸦，瘪嘴
sledgehammer	[ˈsledʒhæmə(r)]	n. 大锤；v. 用大锤打，猛力打
coarse	[kɔːs]	adj. 粗糙的，粗的，粗鲁的，粗俗的，难看的，不雅的
nonabrasive	[ˌnɒnəˈbreɪsɪv]	adj. 未磨损的，不磨损的
shredder	[ˈʃredər]	n. 碎纸机
impede	[ɪmˈpiːd]	v. 阻碍
penetrate	[ˈpenətreɪt]	v. 刺入，穿透，渗入，洞察，了解，被理解，被领悟，看穿
moisture	[ˈmɔɪstʃə(r)]	n. 潮气，水分
component	[kəmˈpəunənt]	n. 部件，组件，成分，分力
pigment	[ˈpɪɡmənt]	n. 色素，颜料；v. 给……染色，呈现颜色
intimate	[ˈɪntɪmət]	adj. 亲密的，密切的；n. 密友，知己；v. 透露，暗示
homogeneous	[ˌhɒməˈdʒiːniəs]	adj. 同种类的，同性质的
diameter	[daɪˈæmɪtə]	n. 直径，放大率
equidimensional	[ekwɪdɪˈmenʃənəl]	adj. 等量纲的，等轴的，等维度的
dimension	[daɪˈmenʃ(ə)n]	n. 大小，尺寸，维度，范围，部分；v. 切削（或制作）成特定尺寸
centimeter	[ˈsentɪmiːtə]	n. 厘米
aperture	[ˈæpətʃə]	n. 孔径，光圈，开口

fraction	[ˈfrækʃn]	n. 分数，小部分，片段，分馏，馏分
stainless	[ˈsteɪnləs]	adj. 不锈的，无污点的，无瑕的；n. 不锈钢
mesh	[meʃ]	n. 网孔，网眼，网
intermediate	[ˌɪntəˈmiːdiət]	n. 中间体；adj. 中间的
screen	[skriːn]	v. 筛查，检查
bronze	[brɒnz]	n. 青铜
nickel	[ˈnɪkl]	n. 镍
monel	[mouˈnel]	n. 蒙乃尔铜-镍合金
agitate	[ˈædʒɪteɪt]	v. 搅拌
gyrate	[dʒaɪˈreɪt]	v. 旋转；adj. 旋涡状的
cylindrical	[səˈlɪndrɪk(ə)l]	adj. 圆柱形的，圆柱体的
pneumatic	[njuːˈmætɪk]	adj. 气动的，充气的，有气胎的；n. 气胎
velocity	[vəˈlɑːsəti]	n. 速度
stationary	[ˈsteɪʃənri]	adj. 不动的，静止的，不变的，稳定的；n. 不动的人或物
compressive	[kəmˈpresɪv]	adj. 压缩的，有压缩力的
incipiently	[ɪnˈsɪpiəntli]	adv. 起初地，早期地
heterogeneity	[ˌhetərədʒəˈniːəti]	n. 异质性，非均匀性
supercritical	[ˌsuːpərˈkrɪtɪkəl]	adj. 超临界的
ozone	[ˈəʊzəʊn]	n. 臭氧
hexane	[ˈheksɪn]	n. 己烷
methylene	[ˈmeθɪlɪn]	n. 亚甲基（methylene chloride 二氯甲烷）
chloride	[ˈklɔːraɪd]	n. 氯化物
pharmaceutical	[ˌfɑːməˈsuːtɪkl]	adj. 制药的，配药的，卖药的；n. 药物
nutraceutical	[ˌnjuːtrəˈsuːtɪkl]	n. 营养制剂，保健品，功能食物
decaffeinate	[diːˈkæfɪneɪt]	v. 脱去……的咖啡因
flavor	[ˈfleɪvə]	n. 风味调料；v. 添加味道
fragrance	[ˈfreɪɡrəns]	n. 香味，香水，香气
lubricant	[ˈluːbrɪkənt]	n. 润滑油，润滑剂；adj. 润滑的，减少摩擦的
fibrous	[ˈfaɪbrəs]	adj. 纤维的，纤维性的，纤维状的
viscosity	[vɪˈskɒsəti]	n. 黏性，黏度
combustion	[kəmˈbʌstʃən]	n. 燃烧，氧化，骚动
analyte	[ˈænəlaɪt]	n. 分析物
reagent	[riˈeɪdʒənt]	n. 试剂，反应物
substantially	[səbˈstænʃəli]	adv. 充分地，大量地，基本上
immiscible	[ɪˈmɪsəbl]	adj. 互不相溶的
intrinsic	[ɪnˈtrɪnzɪk]	adj. 固有的，本质的
rectification	[ˌrektɪfɪˈkeɪʃn]	n. 精馏，整流，修正
viscous	[ˈvɪskəs]	adj. 黏性的，黏滞的，黏的
brine	[braɪn]	n. 盐水

saturated	[ˈsætʃəreɪtɪd]	*adj.*	饱和的，湿透的
condense	[kənˈdens]	*n.*	浓缩
potable	[ˈpəutəbl]	*adj.*	适于饮用的
crystallization	[ˌkrɪstəlaɪˈzeɪʃn]	*n.*	结晶化
uniformity	[ˌjuːnɪˈfɔːməti]	*n.*	均匀性（度），一致
nuisance	[ˈnjuːsns]	*n.*	损害，讨厌的人或东西，麻烦事
effluent	[ˈefluənt]	*n.*	废水，污物，水流；*adj.* 流出的（effluent gas 烟道气，废气）
amenity	[əˈmiːnəti]	*n.*	舒适，便利设施
chamber	[ˈtʃeɪmbə(r)]	*n.*	室腔（settling chamber 沉降室）
centrifugal	[ˌsentrɪˈfjuːg(ə)l]	*adj.*	离心的，远中的；*n.* 离心机，转筒
gravitational	[ˌgrævɪˈteɪʃn(ə)l]	*adj.*	重力的，引力的
tangential	[tænˈdʒenʃ(ə)l]	*adj.*	切线的，正切的，离题的；*n.* 正切，切线
rectangular	[rekˈtæŋgjələ(r)]	*adj.*	长方形的，矩形的，直角的，有直角的
tubular	[ˈtjuːbjələ(r)]	*adj.*	管状的
batch	[bætʃ]	*n.*	间歇，一批
admission	[ədˈmɪʃ(ə)n]	*n.*	允许进入，加入
prevailing	[prɪˈveɪlɪŋ]	*adj.*	占优势的，主要的
fluctuate	[ˈflʌktʃueɪt]	*v.*	变动，上下摇动
pilot	[ˈpaɪlət]	*n.*	飞行员，领航员（pilot plant 试验工厂）
stock	[stɒk]	*n.*	股票，存货，储备
decouple	[diːˈkʌpl]	*v.*	减弱震波，使分离；*n.* 去耦
delivery	[dɪˈlɪvəri]	*n.*	递送，投递，递送物
acetylene	[əˈsetəliːn]	*n.*	乙炔，电石气
arsenic	[ˈɑːsnɪk]	*n.*	砷，砒霜；*adj.* 砷的
sulfur	[ˈsʌlfə]	*n.*	硫黄（sulfuric acid 硫酸）
dichloroethane	[daɪklɔːrəˈeθeɪn]	*n.*	二氯乙烷
grinding	[ˈgraɪndɪŋ]	*adj.*	难熬的，折磨人的，刺耳的
by-product	[ˈbaɪ prɒdʌkt]	*n.*	副产品
purification	[ˌpjuərɪfɪˈkeɪʃ(ə)n]	*n.*	净化，提纯
inventory	[ˈɪnvəntri]	*n.*	详细目录，存货，清单
adjacent	[əˈdʒeɪs(ə)nt]	*adj.*	邻近的，毗连的
sack	[sæk]	*n.*	大袋，麻布袋
carton	[ˈkɑːtn]	*n.*	硬纸盒，纸板箱
bale	[beɪl]	*n.*	大包

Notes

1. The broad implications of this definition have been justified over the past several decades by the types of problems that chemical engineers have solved, although the profession has focused its attention primarily on the chemical process industries.

虽然该专业主要关注的是化工过程工业，但是化学工程师们所解决的问题类型在过去的几十年里证明了这一定义的广泛意义。

2. Chemical engineering developed as a distinct discipline in the twentieth century in response to the needs of a chemical industry that could no longer operate efficiently with manufacturing processes that were, in many cases, simply larger versions of laboratory equipment.
化学工程在二十世纪发展成为一门独特的学科，以应对化学工业的需求，但在许多情况下，化学工业的制造过程已经无法再高效地运作，因为这些制造过程只是实验室设备的简单放大。

3. This approach holds that the unifying concept is not specific processing operations, but rather an understanding of the fundamental phenomena of mass, energy, and momentum transport that are common to all unit operations, and it is argued that a focus on unit operations obscures the similarity of many operations at a fundamental level.
这种方法认为，统一的概念不是具体的处理操作，而是对所有单元操作所共有的质量、能量和动量传输的基本现象的理解，有人认为对单元操作的关注掩盖了许多操作在基本层面上的相似性。

4. In solving rate problems, such as heat or mass transfer, with this simple concept, the major difficulty is the evaluation of the resistance terms, which are usually calculated from an empirical correlation of many determinations of transfer rates under different conditions.
在利用这个简单的概念解决传热或传质等速率问题时，主要的困难在于阻力项的评估，通常根据不同条件下许多传递速率决定因素的经验关联来进行计算。

5. In addition, the properties of a material can be significantly affected by particle size; for example, the chemical reactivity of fine particles is greater than that of coarse particles, and the color and hiding power of a pigment is significantly affected by particle size.
此外，材料的特性也会受到颗粒大小的显著影响；例如，细颗粒的化学反应活性要高于粗颗粒，颜料的颜色和遮盖力也会受到颗粒大小的显著影响。

6. Although the continuous process requires more carefully designed equipment than the batch process, it usually requires less space, is more easily integrated with other continuous steps, and can be operated at any prevailing pressure with venting to atmospheric pressure.
虽然与间歇式工艺相比，连续式工艺需要更精心设计的设备，但它通常只需更小的空间，更容易与其他连续步骤集成，并且可以在任何通行压力下运行，并可排放至大气。

Trivial — Last but not the Least

The Road to Graduate School: A Guide to the Application Process

When it's time to apply to the Graduate School of Chemistry and Chemical Engineering for your master's (M.S.), professional science master's (P.S.M.), or doctoral (Ph.D.) degree, the next important steps include the following (1) selecting an appropriate program, (2) taking your Graduate Record Examinations (GREs), (3) completing some application forms (including preparing a tailored personal statement), and (4) submitting your applications along with your transcripts and some letters of recommendation. You should begin preparing for this process about a year before the application deadline, since most deadlines are between December and February.

Finance management

There is a fee for applying to graduate school. The cost to take the GRE General Test is just over $200, while the GRE Subject Test costs an additional $150. You can also expect to pay for an official copy of your transcript. In addition, most graduate programs have application fees. If you apply to six or seven programs, it can easily cost you $700. Sometimes, applying for the GRE Fee Reduction Program can offer a 50% discount on exam fees in certain situations, and at the same time, some graduate programs may waive application fees. Therefore, checking with the Graduate Admissions Office to see if you qualify may help you save some money.

In addition, you should know that financial aid is a typical choice for chemistry graduate students. Most programs offer tuition discounts and living stipends for students. The stipends are usually exchanged for the first or second year of teaching and subsequent research. Therefore, the most important financial expenses when applying to graduate programs in chemistry and chemical engineering are exam and application fees.

Qualify for the criteria

There is no one-size-fits-all application for graduate programs. Each has its own processes, requirements, and deadlines. Before applying to a graduate program, you must carefully review each program's criteria to ensure that you tailor your application plan to the institution's application requirements as much as possible while still meeting the criteria. Make sure that you are not rejected from the project due to technical issues.

The following are the materials that most applicants will need to prepare:

Transcripts

Generally, the programs to which you are applying will require official copies of transcripts from each institution you have attended since high school. Make sure you have all your transcripts ready to submit before the application deadline; your application will be considered incomplete without them.

GREs

The GRE General Test includes Verbal Reasoning (two sections), Quantitative Reasoning (two sections), and Analytical Writing (one section). The entire test takes four to five hours to complete (including breaks). If you are required to take the GRE Subject Test, it takes approximately one and a half to two hours to complete and covers all four years of your undergraduate program, including your senior year. Therefore, it is helpful to take the Subject Test later in the academic year. But don't wait too long, because the scores must be submitted with the rest of your application before the application deadline.

You can get help from the EST, which administers the GREs, and has information, test times and locations, and — best of all — practice tests on its website. You can also find workshops and study guides to help you prepare.

Personal statement

The graduate school application process requires several personal statements, and you need to take full advantage of these statements to showcase your accomplishments and grades. First, it is important to understand the requirements of the graduate program. Tell the graduate admission officer who you are and how you will contribute to science and the world. In addition, your personal statement can highlight research, leadership, and other experiences that will impact your career. Since research is the cornerstone of most chemistry programs, describing your undergraduate research and what you learned from it will help you make your case. You should mention the specific department you are interested in and why. This will help readers visualize your presence in school and show your interest in the project. Please remember to proofread your statement.

Letters of recommendation

Graduate programs generally require three letters of recommendation. These should be from people who know you in a way that is relevant to the program. For example, a doctoral program is based on research, so the professor who supervised your undergraduate research must provide a letter. Letters can also come from your advisor, a professor who teaches the courses in which you excel, or the teacher for whom you serve as a teaching assistant or performance instructor. Please notify your recommender as soon as possible and provide as much detail as possible about the letter — deadline, submission information, name of recipient, topic to be addressed, and so on. Your personal statement is very helpful to the writer.

In short, you can ask your fellow graduate students for case studies of their application materials, and you can ask them to proofread your personal statement. Graduate school application risk is high, but as thousands of graduate students have proven, it is manageable. With a little determination, a little planning, and a lot of proofreading, you can be one of them!

Part C
English for Basic Academic Communication

Unit 10

International Conference

10.1 An Overview of Academic Communication

Academic communication, also called scholarly communication, refers to highly structured methods of communication that are generally used only in educational settings. Academic communication can include the words and structures used to express ideas, as well as the methods by which ideas are disseminated. Many people intuitively know the proper way to communicate in different settings, but the way you talk to your friends is usually different from the way you talk to your peers. As a student of chemistry and/or chemical engineering in the 21st century, you must feel increasingly required to participate in many professional and academic-related activities, such as attending international conferences, submitting a scientific manuscript, communicating with relevant laboratories outside China, searching for international jobs, etc. which are effective ways to broaden your exposure to academic circles and become actively involved in scientific activities in the world.

Academic communication is a basic skill for people who have studied chemistry and/or chemical engineering that need to learn. Many schools offer courses designed to teach or improve these communication skills. In these courses, students learn advanced vocabulary, proper grammar, reading comprehension, and verbal expression, as well as how to write in a formal tone and address the reader politely, whether it's a professor or a peer. In addition to developing high-level communication skills, the courses also cover academic-specific topics, such as how to format a paper according to Modern Language Association (MLA) standards and how to write book reviews, research papers, and lab reports. In college, however, writing your thesis is the first element of academic communication.

People with English as a Second Language (ESL) may require intensive instruction in academic communication if they have special learning, speech, and/or hearing needs. Special learning techniques are often used to ensure that students learn to communicate academically as easily as possible. For example, in the case of ESL students, this may mean extra practice with pronunciation or complex words used in academic argumentation.

Academic communication is the communication of academic information. When a scholar needs to publish a new discovery, or your teacher needs to tell you about resources to use for a homework assignment, they need to use a specific communication channel. In these cases, scholarly communication refers to the methods by which ideas are disseminated among scholars or students. You can find scholarly information disseminated in places like academic journals, newsgroups, online course management systems, and research libraries.

Today, the widespread use of the Internet has had a profound impact on scholarly communication, bringing both advances and complications to academic discourse. While academics now have the ability to share information with all corners of the globe, they also face an increased risk of plagiarism. For this reason, proper citation of sources through annotations or bibliographies is essential to academic communication. At the same time, academic institutions need to keep up with changing technology and may need to change their policies to ensure fair access to academic resources. They may also need to adapt to new forms of academic communication, such as the use of email to communicate with instructors or the need to cite online articles.

10.2 International Conference

As a future practitioner in the chemical industry, you may have the opportunity to attend a variety of meetings, the regional and/or webinars, both large and small, such as conferences, symposia, congresses, colloquia, conventions, convocations, annual meetings, biennial meetings, meetings held every three years or more, forums, summits, seminars, workshops, roundtables, special panels, poster sessions, exhibits, expositions, or other similar situations. Your success in attending a meeting depends on how effective you are. This section is designed to help you understand the important steps to prepare for attending a meeting and to provide you with a variety of skills and techniques to help you be effective. It will also provide you with the knowledge, practical skills, and useful suggestions to help you be successful when attending international meetings.

10.2.1 Sources for Conference Information

It is beneficial to attend a professional meeting in your field, whether you present a paper or not. It will be good for you to read the information about or the proceedings of a meeting, even if you do not have the opportunity to attend. Attending professional meetings and reading conference literature are good ways to keep abreast of the latest developments in your field, which should become an important part of your professional agenda.

(1) The Internet

The Internet is a popular source of information today, with over a billion pages available every day. Both Internet search engines and meta-search engines are powerful tools for those who want to participate in some professional activities. Internet search engines are tools that use computer programs to automatically gather information from the Internet. With this information, they create a database. Each of the major search engines tries to do the same thing, which is to index the entire Web, so they handle a huge amount of data that is updated frequently. Search engines provide access to very large collections and provide the most comprehensive search results.

Search engines are the best tools to use when you are looking for very specific information or when you are looking for possible meetings in your field. If you need information on a very detailed or multifaceted topic, a search engine will usually give you not only more information, but also the most accurate and up-to-date information. If you are looking for a specific meeting or possible upcoming meetings, a search engine is the best place to start. And you would be wise to use more than one, because each engine will return different results, because although most of the major search engines attempt to index the entire Web, each has a different way of determining which pages are most relevant to your search query. In one search engine's database, a relevant document may be the 100th on the list; in another database, the same document may be the first. To find the most relevant documents, you should become familiar with many search engines and their features, and learn to use the Internet effectively and efficiently to make the most of the resources and services available online.

If you are looking for information about a specific meeting, you can use the name of the meeting as the selected keywords; if you are looking for some possible upcoming meetings, you can use words that describe the important concepts of your search, such as "nanotechnology" "electrochemistry", plus one of the following keywords: meeting, conference, congress, convention, symposium, forum, call for papers, call for abstracts, announcement, activities, calendar, and so on. Or you can enter a year, such as 2023, and hope to find something that takes place in the year you select.

If the results are not satisfactory, try the same keywords in a different search engine. Some search engines support concept searching; that is, the search engine automatically searches for synonyms of the words you enter. For example, if you search for "symposium", it may also return results with the word "workshop" or "conference". When using the Internet, remember that you are sharing a resource that is distributed around the world. There will be times when the network is not working perfectly. In any case, be persistent and willing to learn new things.

(2) Professional journals, membership of professional organizations, professional groups, and conference literature

In addition to the Internet, professional journals, membership in professional organizations, professional groups, and conference literature are good places to look for conference information. In a journal, you can often find a few newsletters, announcements, or calls for papers, as well as calls for abstracts, posters, or proposals for specific meetings in your field. Membership in professional organizations, such as societies, associations, and federations, allows you to enjoy the privileges of sharing information and participating in the organization's activities, and sometimes you can register for meetings as a member at a reduced rate or for free. For example, as a member of the Chinese Chemical Society, you can easily get all the information about the regional conference in the coming year.

Most professional organizations are open to all scientists and students who wish to participate and share their goals. You can easily find the information of various organizations on the Internet. In addition, the conference information can also be conveniently received in many professional groups and/or official account via mobile phone APP, such as Wechat.

Conference literature is also a good source of information about future conferences. You will be well informed about the latest developments reported at past professional meetings in your field if you pay close attention to the conference literature printed during the meeting, or printed before or after the meeting[1]. In the meantime, you can also learn about upcoming activities by paying close attention to the conference literature published in your research area. There are various forms of conference literature, such as preprints, advance abstracts, pre-abstracts, abstract books, proceedings, symposia, transactions, reports, records, prints, papers, etc. Besides the conventionally printed conference literature, you can also find the e-journal online for the same information. It is easy for you to get a lot of useful news which is very helpful for you to keep in step with the recent development in this field. And you can also get some papers from the website, since their copyright policy is as follows: "In general, each conference retains the copyright to its proceedings, but allows copies of individual papers to be taken from this site, provided that the source is acknowledged".

10.2.2 Conference Organizers and Session Modes

Different conferences are organized in different ways by different sponsors and organizers. For some conferences, there is only one secretariat or program committee in charge of everything related to the convening of the conference, while for other meetings, several committees are established to take care of different tasks. In order to attend a meeting, you need to get enough information about the organizing committees and the way the conference is convened.

(1) Organizers and committees

If you need more information or have questions about the meeting you are attending, you can contact an appropriate committee for the meeting. Typically, one or more of the following committees will be established to call a meeting.

① Organizer and sponsor. Each meeting must have an organizer, who is the person or organization responsible for managing the event and, in conjunction with a sponsor, determining the topic or topics to be presented. It is the organizer's responsibility to select a venue, assemble a guest speaker or panel of guest speakers, formulate an invitation to interested parties, appoint a chair, and hire the necessary staff to run the event. The organizer is the primary agent who should plan everything for the meeting and instruct the selected chair, secretary, and support staff in their required duties. Therefore, if you need to clarify or confirm anything, you should contact the organizer, not the sponsor.

② The secretary. One of the major responsibilities of the secretary is to distribute the program and publish the preprints and proceedings of papers or abstracts presented. Since the secretary's name appears on all publicity for a conference, this is the first person to contact and the person to complain to if necessary. It is also the secretary's job to introduce the chair to the audience. He or she is also responsible for taking minutes and distributing them to all organizing committee members.

③ Organizing committee/program committee. The organizing/program committee has overall responsibility for decisions made for the success of the conference. They work on issues related to the quality of the meeting, including key participants, funding sources, and relevant topics; receive papers, abstracts, and posters if there is no scientific committee or paper committee; select session topics and chairs, sort abstracts, organize sessions, and hand off to session chairs; be responsible for assembling the program and creating a database of authors with their addresses and titles; ensure the financial solvency of the meeting, check anticipated expenses & projected revenues, corporate sponsorships, and registration and accommodation fees; and secure funding sources, including coffee break sponsors, program sponsors, banquet sponsors, workshop sponsors, etc., if there is no finance committee.

④ Logistics committee. The logistics committee coordinates all meeting and presentation facilities, food services, registration, presentation equipment, room accommodations, and other general conference logistics required to host the conference at the selected site. They set up programs for spouses and other companions during the conference if there is no companion committee.

⑤ Session organization chair/convener. The person in this role will work with the session chairs to solicit papers for the session, ensure authors submit abstracts on time, run the session on time; work with the technical program chair to notify authors of manuscript acceptance/rejection, schedule, and proper formats; and arrange all speakers and programs for the conference.

(2) Session modes

Conference papers are typically presented in a variety of modes to encourage maximum interaction between authors and attendees. Final decisions on presentation mode are made at the time of abstract or paper acceptance. Sometimes a small portion of the poster and oral session space is reserved for the presentation of late-breaking developments. These are selected from proposals submitted to the program committee a few weeks or days before the meeting, or at the meeting. The presentation formats described below are intended as a guide to possible ways to present your work at a conference. Presenters at different meetings are generally encouraged to explore creative and innovative alternative formats, especially those that invite audience participation and engagement.

① Keynote sessions. Keynote addresses are usually given by well-known and experienced specialists in the field. Keynote speakers and session chairs would set the tone for or support the theme of the conference. These presentations are by invitation only. They are presented to the entire conference audience in plenary or keynote sessions, and are usually included as full-length papers in the conference proceedings. Although these papers are by invitation, neither the presentation time nor the length of the paper should exceed the specified limit.

② Oral sessions. The core of the conference program is the oral presentation of peer-reviewed, full-length papers. They will be organized in plenary sessions, panel sessions, or concurrent sessions, each consisting of several papers. These sessions provide an opportunity for more in-depth discussions between authors and participants. Full-length papers should not exceed certain time and page limits. These papers will be the main part of the conference preprints or proceedings.

③ Poster sessions. Abstracts or papers that receive high scores are usually scheduled for oral presentation in plenary or concurrent sessions, while others are scheduled for poster sessions. However, at some conferences, you may be invited to present your work as a poster in addition to the oral presentation. Poster sessions are a valuable way for authors to present their work and meet with interested viewers for in-depth technical discussions. These papers are also published in the conference proceedings as abstracts or full-length papers.

④ Research-in-progress sessions. Some conferences reserve a time slot for research-in-progress sessions. Research-in-progress papers can be submitted as short papers or abstracts describing an ongoing research project or study. Sessions typically include a chair and several presenters who offer their perspectives on a single topic.

⑤ Discussion/question & answer session. On most occasions at academic conferences, the oral presentation is immediately followed by a discussion or Q & A session, sometimes a one-on-one Q & A session, sometimes a general discussion after several papers have been presented.

(3) Session format

For different conferences, sessions are organized in different formats. In general, session titles should be kept relatively general so that papers can easily fit into sessions that make sense. Here are some examples of different session formats as follows.

◇ Full Paper sessions will be organized into 105-minute time blocks with no more than six (6) papers per session. Each presenter will have 15-20 minutes including time for questions/discussion. Keynote sessions will be organized into 50-minute time blocks with no more than two (2) papers per session. Delegates with special time requirements should contact the technical program chair. Presenters are asked to check in with the session chair no later than 30 minutes prior to their presentation. Neither presentation time nor paper length should exceed the specified limit for each session, including invited keynote addresses.

◇ In a ROUNDTABLE format, the moderator introduces the topic with approximately three minutes of overview and scene-setting information. The moderator briefly introduces each panelist and puts each in perspective in relation to the session topic. Each panelist gives a brief five to seven-minute background presentation that develops the panelist's point of view. The moderator asks questions, generates discussion, engages the audience, and tells the story through the interaction of all participants. In the final two or three minutes, the moderator summarizes the discussion, highlights key points of agreement and disagreement, identifies resources for further study, and ends the session on time. In a PRESENTATION PANEL format, the moderator introduces each speaker and puts each presentation in perspective. Each speaker gives a prepared presentation of 15 or 20 minutes. The moderator asks questions and elicits responses from the audience.

◇ Two formats have traditionally prevailed at our conventions. In the first, a keynote speaker speaks for about 30 minutes, followed by a 15-minute commentary and a number of shorter statements of opinion, with the remainder of the time devoted to general discussion. An alternative format involves three 20-minute papers on different aspects of the topic, followed by general discussion.

◇ The exhibitor theater will be available on Monday, Tuesday, Wednesday, and Thursday from 11:30 a.m. to 1:45 p.m. Time will be allocated to companies in one-hour blocks on a first-come, first-served basis. Each company may request one or two hours. Sessions will be scheduled from 11:30 a.m. to 12:30 p.m. and from 12:45 p.m. to 1:45 p.m. each day. Each one-hour session should include time for Q & A.

◇ The four-day conference will include plenary, oral, and poster sessions on research in plasma science and technology, and a keynote address by the recipient of the Plasma Science and Applications Committee Award. The oral sessions will include both invited and contributed papers. Invited papers will be allotted 25 minutes for presentation and 15 minutes for questions. Contributed papers will be allotted 12 minutes for presentation and 5 minutes for questions.

◇ All presentations are 12 minutes in length unless otherwise noted in the meeting program (8 minutes for presentation; 4 minutes for Q & A). THIS IS AN ABSOLUTE LIMIT!

◇ None of our technical sessions compete with exhibit hall hours, and all sessions run in a single-track format, so you do not have to sacrifice one session for another.

10.2.3 Papers, Abstracts, Posters and Proposals Submitted to Conference

Papers, abstracts, posters, and proposals are required to attend most conferences in the academic fields. You can be invited only if your submission is accepted by the conference organizer. Therefore, if you have a good idea of how to prepare an acceptable paper, abstract, poster, or proposal, you will have a greater chance of successfully receiving an invitation from the conference you are interested in.

(1) Papers

Papers submitted to a conference must follow certain rules and styles similar to those required for a journal paper. If you do not follow these requirements, your paper may be rejected simply because of formatting issues before it is presented to your peers for review. Here is a good example of a detailed paper style that reflects the requirements for papers submitted to an academic symposium. You can get a quick idea of some common practices for conference papers if you can understand every detail in these instructions.

Sample of the "Instructions to Authors of Papers" for the "38th U.S. Symposium on Rock Mechanics".

Authors of Proceedings papers must type them in a form suitable for direct photographic reproduction by the publisher. In order to ensure a uniform style throughout the volume, all papers must be prepared strictly according to the instructions provided by the organizer. A laser printer should be used to print the text. The complete camera-ready copy will be reduced to 75% by the publisher and printed in black only.

1 GENERAL INSTRUCTIONS

1.1 Type Area

Print the text in black on white paper. Use only one side of the paper. The text should fit exactly into the type area (187×270 mm). The width of each column should be 90 mm, no more and no less. Leave a 7 mm gutter between each column. The total width of the type area should be exactly 187 mm and the height of the text on each page should be exactly 270 mm.

1.2 Typeface, Font Size, and Indentations

Use "Times New Roman" 12 point size with 13 point line spacing. Do not use a sans-serif font such as Univers, Helvetica, or Gothic. Use Roman type except for headings, parameters in mathematics, Latin names of species and genera in botany and zoology, and titles of journals and books, all of which should be in italics. Never use bold, except to denote vectors in mathematics. Never underline text.

All text should be justified to a column width of 90 mm. Indent the first line of each paragraph by one em-quad. Do not indent after an open line or a heading.

2 LAYOUT OF TEXT

2.1 Title of paper

Type the title of your paper in lower case at the top of the first page. It should be no longer than 70 characters.

Type the name(s) of the author(s) below the title. Initials should precede the last name. The affiliation of the author(s) should follow in the next line. These texts will be retyped by the publisher.

Write a tentative page number and the name(s) of the author(s) at the bottom of each page, outside the type area.

2.2 Abstract

Each paper should begin with an abstract of no more than 150 words. Type the first line of the abstract on the 17th line 73 mm from the top of the typo area, as shown on the transparent typo area overlay. Type the word ABSTRACT followed directly by the abstract itself, which should run over both columns. Leave two blank lines before the text (or first heading) of the paper begins.

2.3 Headings

Type primary headings in Roman capital letters and secondary headings in lowercase. Type headings flush with the left margin of the appropriate column. Leave two blank lines above and one below the primary headings and 1.5 blank lines above and 1/2 blank line below the secondary headings. Do not indent the heading or the first line of text after the heading. If a heading is at the bottom of a column, move it to the top of the next column, leaving space at the bottom.

2.4 Photographs, figures, equations and tables

Number figures consecutively in the order in which they are referred to in the text, making no distinction between diagrams and photographs. Figures may be either 90 mm wide (one column) or 187 mm wide (two columns).

Figures, photographs etc. should be in black only and should be pasted onto the typescript where you want them to appear in the text. Do not place them sideways on a page; however, if this cannot be avoided, no other text should appear on that page. Figures, etc., should not be centered, but should be placed against the left margin. Leave about two lines of space between the text and the figure (including the caption). Never set any text in the same column next to a figure, table, or photograph. The most convenient place for these items is at the top of the page. When pasting the figures, remember to place the actual lines of the figure immediately against the top of the type area: ignore the open space which may be present above the lines of the figure[2]. Diagrams may also be printed from file into the text. Line drawings (as well as photographic reproductions of these) should be in black on white. Keep in mind that everything will be reduced to 75 percent. Therefore, 2 mm should be the minimum size of the lettering. Lines should preferably be 0.2 mm thick. Keep figures as simple as possible. Avoid excessive notes and labels.

Photographs should be black and white, with good contrast and on glossy paper. Photographic reproductions cut from books or journals, photocopies of photographs, and screened photographs are unacceptable. Do not scan photographs.

Set equations right against the left margin of the column and number them consecutively. Refer to equations by these numbers. Leave one blank line between equations and text, and between two equations.

Number tables consecutively and locate them close to the first reference to them in the text. Avoid abbreviations in column headings. Indicate units in the line immediately below the heading. Explanations should be given at the foot of the table, not within the table itself. Use only horizontal lines; align all headings to the left of their column, and start these headings with an initial capital. Type the caption above the table to the same width as the table.

3 REFERENCES, SYMBOLS AND UNITS

Consistency of style is very important. Note the spacing, punctuation and caps in all the examples below.

—References in text: Figure 1, Figures 2-4, 6, 8a, b (not abbreviated)

— References between brackets: (Fig. 1), i (Fig. 2-4, 6, 8a, b) (abbreviated)

—USA/UK instead of U.S.A./U.K.

—Author & Author (1989) instead of Author and Author (1989)

—(Author 1989a, b, Author & Author 1987) instead of (Author, 1989a, b; Author and Author, 1987)

—(Author et al.1989) instead of (Author, Author & Author 1989)

Always use the official SI notation:

—kg/m/kJ/cm instead of kg. (kg)/m. kJ. (KJ)/cm.

—9000 instead of 9,000, but if more than 10,000: 10,000 instead of 10000

4 NOTES

These should be avoided. Insert the information into the text. In tables, the actual footnotes are placed directly below the table.

5 CONCLUSIONS AND REFERENCES

Conclusions should state concisely the most important propositions of the paper as well as the author's views of the practical implications of the results.

In the text, place the authors' surnames (without initials) and the date of publication in parentheses. At the end of the paper, list all references in alphabetical order

(2) Abstracts

When you prepare and abstract for a paper to be presented at a conference, you should also follow the rules for writing a journal paper abstract. You should state the purpose of the study, describe the methods, summarize the results in sufficient detail to support the conclusions, and give the conclusions reached. For a conference abstract, statements such as "The results will be discussed" or "Further data will be presented" are unacceptable. Simple tables or graphs may be included if they fit within the given abstract form. When using abbreviations, spell out in full at the first mention, followed by the abbreviation in parentheses. Make the title as specific as possible.

(3) Posters

A poster is simply a static, visual medium (usually of the paper and board variety) that you use to communicate ideas and messages. The difference between a poster and an oral presentation is that you should let your poster do most of the "talking", i.e., the material presented should convey the essence of your message. However, this does not mean that you can completely disappear from the poster wall. You need to "stand by" your poster as often as possible, when time is available, to answer questions and provide further details.

There are several advantages to poster presentations. The purpose of poster sessions is to allow for the mutual exchange of ideas and data in a more informal, comfortable setting. A poster can allow for a clearer and more reflective process of presentation, especially of statistical or visual information. The narrative of each poster, including key points and conclusions, can be highlighted and thus absorbed at each person's own pace while viewing the poster. In addition, posters promote international exchange of ideas and research by providing a format for non-native speakers of the working language to optimally present their research with greater clarity and understanding. Posters are typically highlighted during the meeting by being placed in conference venues and high activity areas associated with coffee breaks, exhibits, or other activities.

As with an oral presentation, there is a standard format for a poster. They are usually divided into the following sections:

✧ A title section that tells others the title of the project, the people involved in the work, and their affiliations;

✧ A summary of the project, stating what you set out to do, how you did it, the key findings, and the main results;

✧ An introduction, which should include clear statements about the problem you are trying to solve, the properties you are trying to discover, or the proofs you are trying to establish. These should then lead to statements of project goals;

✧ A theory or methodology section that explains the basis for the technique you are using or the approach you have taken in your study. You should also state and justify any assumptions you make so that your results can be viewed in the proper context;

✧ A results section that provides illustrative examples of the main findings of the paper;

✧ A conclusions section that lists the main findings of your research;

✧ A further work section, which should include your recommendations and thoughts on how the work could be continued, other tests that could be applied, etc.

Try to design your poster carefully, using color or number coding or circles and lines to help you identify and categorize information. Ultimately, poster design is a personal matter, and different people will have different ideas about how best to present certain information. However, here are some general rules to follow:

✧ Keep the material simple. You should make full use of space, but do not cram a page full of information, as the result can often look cluttered; be concise and do not ramble; use only pertinent information to convey your message; be selective in presenting results. Present only those that illustrate the main findings of the research. However, keep other findings handy so that you can refer to them if asked.

✧ Use color sparingly and tastefully. Colors should be used only for emphasis, differentiation, and interest. Do not use color just to impress. Try to avoid using large swaths of bright, garish colors such as bright green, pink, orange, or purple. It is better to keep the background light because people are used to it. Choose background and foreground color combinations that have high contrast and complement each other, such as dark blue on white or very light gray. Avoid using gradient fills. They may look great on a computer screen, but unless you have access to a high-resolution printer, they can look really messy on paper.

✧ Use no more than two fonts. Too many fonts are distracting, especially when they appear in the same sentence. Fonts that are easy on the eyes are Times New Roman and Arial. Titles and headings should appear larger than other text, but not too large. Text should also be legible from a distance, such as 1.5 to 2 meters.

◆ Do not use all capital letters in your posters. Capital letters can make the material difficult to read.

◆ Use pictures, graphs, and charts correctly. A picture is worth a thousand words, but only if it is properly drawn and used. Choose graphic types that are appropriate for the information you want to present. Annotations should be large enough, and the lines of line graphs should be thick enough to be seen from a distance. Instead of using lines of different thicknesses, use contrasting colored lines or different line styles to distinguish different lines in multi-line charts. Graphs and drawings should be labeled and should be large and clear enough to be read from a distance.

◆ Maintain a consistent style. Inconsistent styles create an impression of disharmony and can interrupt the fluidity and flow of your messages. Headings on different pages of the poster should appear in the same position on all pages. Graphics should be the same size and scale, especially if they are to be compared. If bold type is used for emphasis on one page, do not use italics on other pages. Captions for graphs, drawings, and tables should be placed either at the top or bottom of the figure.

In general, if you are presenting your poster at a conference, you will have limited space. The amount of space you are allowed will determine the content of your poster. Find out how much space you will be given and take into account the main criticisms of posters in the past, such as

◆ Inability to read the poster from a distance of 1.5m or more due to small letters, numbers and symbols;

◆ Too much information and text;

◆ Objectives and conclusions that are not clearly stated;

◆ Figures and labels that are not self-explanatory.

Therefore, in order to effectively communicate the results of your research in the relatively short time available to viewers, you should put considerable effort into the design of your poster. Pay attention to the specific details outlined in the instructions provided by the organizing committee of the conference you are attending.

One last piece of advice: check your spelling. There is nothing more amusing or annoying than spelling mistakes in public, especially on the title page. Spelling mistakes give the impression that you are not serious and professional about your work, that you have not put enough effort into it, and that your work is not worthy of high scores.

Study the poster preparation tips in the following example:

Sample of the "Poster Presentation Instructions of the 2018 IEEE International Ultrasonics Symposium"
One poster board is allocated to each presentation. The recommended poster size is **Landscape format**, Arch E which is 36 inches high by 48 inches wide (92 cm×122 cm). European alternative ISO A0 (84 cm×119 cm) can also be used. Posters must be mounted using push pins provided by the organizing committee. Each poster presenter is required to defend his/her poster during the respective poster

session slot for the paper to be included in the conference proceedings.

Simply posting the pages of your written version of the proceedings paper is NOT effective and thus NOT acceptable for your poster.

The title of your poster should be done in block letters which are AT LEAST 8 to 10 cm (3 to 4 inches) high.

All text must be easily readable from a distance of 1 to 2 meters. Make the lettering at least 1 cm high, smaller lettering will not be legible from a distance of 1 to 2 meters.

All graphs and charts should be AT LEAST 25×30 cm (approximately 8.5×11 inches) or larger.

It is a good idea to sequentially number your materials in the poster. This will indicate to the viewer a logical progression through your poster.

Provide an introduction (outline) and a summary or conclusion for your poster.

Prepare your poster carefully so that it can be used as the basis to explain and answer questions from the viewers.

It is helpful to have copies of the written version of your paper available for those viewers who may want to study specifics of your work in more detail.

Have your business cards available for those who may wish to contact you at a later date.

Bring along a tablet of blank paper that you may use for a discussion of technical details relating to your poster.

(4) Proposals

Some conferences invite you to organize and chair panel sessions. Typically, these proposals include a cover sheet with the name of the panel and the names, affiliations, addresses, phone and fax numbers, and email addresses of the organizer and all presenters, as well as a general summary followed by one-page abstracts of each paper.

We do not discuss proposal-related sessions here because of the limited involvement of rank-and-file chemical workers.

(5) Review and acceptance of conference papers and abstracts

All papers, abstracts, posters and proposals will be reviewed with certain criterial by the conference organizers. The standards are different, although some essential criteria will be similar. The following aspect should be paid attention to:

- ✧ The subject
- ✧ Interpretation of results
- ✧ New and original contribution
- ✧ Organization and length
- ✧ Clarity
- ✧ Suggestions
- ✧ Illustrations and tables
- ✧ Title, topic and keywords
- ✧ References
- ✧ Acceptance qualifications

10.2.4 Financial Assistance for Attending Meetings

In addition to the quality of your work, financial considerations will always be a factor in your decision to attend. Expenses to consider for attending a conference include transportation (airfare, transfers, taxi, shuttle, subway, streetcar, etc.), conference fee (registration, materials, or social programs, other recreational activities, etc.), lodging (hotel room and meals), taxes, tips, incidental expenses, etc.

If it is difficult for you to raise the funds to attend a conference you are very interested in, you will need to raise funds to support your visit. Some possible sources of financial support are: membership, student scholarships, special awards, early registration, funds from foundations and other organizations such as your own institution, the National Foundation of Sciences, and your foreign or domestic collaborative research partner. If you make some positive attempts, you may be able to obtain full financial support for your participation, or at least justify a larger grant than others. Always remember that there is a chance that you will receive financial support without any effort on your part.

(1) Membership and studentship

As we learned earlier in the International Conferences Information Unit, membership is a very good source of financial support for attending conferences. However, if you are not a member of a scholarly organization, you can take advantage of your student status. To encourage student participation, some conferences offer a limited number of travel grants and a waived or reduced registration fee. You can apply for this support if you are a graduate (doctoral or master's) or undergraduate (bachelor's) student. Sometimes graduate students are especially encouraged to apply. The Student Travel Assistance Grant is often intended to provide partial reimbursement for lodging and transportation expenses. The stipend will be paid either in advance of the conference or as a reimbursement, in which case you will pay for travel expenses and then submit a voucher and original hotel and transportation receipts documenting those expenses[3]. It is expected that you or your institution will be responsible for any travel expenses not covered by the scholarship. To apply for the scholarship, you will need to complete and submit several forms, which are often available on the conference website or at the conference headquarters. Typically, you will need to complete an application with a statement of your current research interest and its relevance to the conference, and obtain a letter of endorsement from a faculty member, your dean, or your advisor at your institution. The letter of endorsement should address (a) the educational benefit you will gain from attending this conference; (b) your potential for future contributions to the field; (c) a commitment that expenses not covered by the stipend will be paid by your university or by you personally.

The following list of the registration fee charged at the 45th PIERS (Photonics and Electromagnetics Research Symposium) held in Chengdu, China is a good example to illustrate the privilege that a student can enjoy in the aspect of fees :

Registration fee	Onsite participants
Early students fee (before 15 January, 2024)	USD 380 / RMB 2660
Students fee (after 15 January, 2024)	USD 490 / RMB 3430
Early regular fee (before 15 January, 2024)	USD 620 / RMB 4340
Regular fee (after 15 January, 2024)	USD 730 / RMB 5110

(2) Funds from the conference sponsor

Often the sponsor of a conference will have a limited amount of funding or scholarships available to support attendees, and these are taken into account when the conference is announced. If you wish to apply for funding to support your attendance at the conference, you should do so as soon as possible, taking into account the time needed to write and process your application, and the deadlines for applying for funding. In most cases, you will be eligible for a reduced fee if you register early.

(3) Awards

Some conferences present special awards, such as "Best Paper Award" "Best Student Paper Award", or other awards. If you have prepared a high quality paper, abstract or poster, you can try to apply for these awards. If you could win these awards, it would not only be a matter of scholarships or certificates, but it would also contribute to your academic development and scientific reputation.

Expect that a single sponsor will not be able to cover the entire cost of your conference attendance. They are likely to contribute to a single aspect, such as travel or the conference fee. Therefore, when applying for funding, think broadly about the possible funding channels and focus on the elements that meet the conditions of the funding, e.g. are you presenting a paper at the conference, are you a member of an organization, are you a student, will you be competing for a possible award, will your institution or advisor support you, will you put things learned or exchanged at the conference into practice for the benefit of your ongoing research project/your university/your community, etc., which happens to fall within the purpose of some foundations or organizations. It would be great if you could get more than one funding agency to support your participation.

10.2.5 Rehearsal, Attendance and Culture Considerations

Once you receive an invitation, you will need to prepare yourself. The ability to present and speak well in front of your peers is a critical professional skill that can contribute to the success of your academic career. Whether you are giving a keynote conference talk, a session presentation, or answering questions in a discussion, you want to be heard and get your message across. To achieve this goal, you will need to practice and rehearse your presentations many times, especially if you are new to public speaking. Even if you are an experienced speaker, rehearsing ahead of time can help you be more successful.

(1) Rehearsal and culture considerations

For an effective performance, you need to rehearse a lot, because a good speech with a good delivery style takes practice. It is worth a lot of effort and work to develop your confidence and platform speaking skills. Another basic requirement for an effective presentation is to address the background and interests of the audience, such as cultural values and attitudes, with appropriate oral, written and body language. And a well-prepared mind for multicultural backgrounds will enable you to be more successful in communicating with your colleagues from other countries before and after the conference.

① Pre-conference rehearsal. With more rehearsals and more public speaking experience, you will gradually become more comfortable and effective with groups of any size, anywhere, any time. The suggestions in this section are designed to help you make your presentation look and sound more professional. The best way to become an excellent presenter is to watch really good, experienced speakers and model your talks after theirs. Notice not only what they say, but also what they do: how they structure their talks, how they move, how they sound. Add these elements to your own presentation. Plan to rehearse several times, alone and with trusted friends or colleagues who can listen to your presentation and give you honest feedback to help you improve. The main purpose of rehearsing is to find out how you sound. If possible, record your speech and play it back. Listen for words, especially the specific terms that are crucial to your presentation, to see if your pronunciation is difficult to understand, and then practice pronouncing the difficult word that sticks in your throat correctly, or replace such words with ones that come through loud and clear.

Also, practice varying your pitch and intonation. Instead of speaking in a monotone, let your voice emphasize key points. Look in a mirror to observe your gestures, posture, and facial expressions. Use your hands only when you need to emphasize a point. Pace your speech according to the familiarity of your topic. Slow down when you are introducing something new.

After rehearsing alone, practice formally in front of your staff at a formal meeting. Rehearsing with an audience will help you determine how the audience is responding, where to add another visual, and when to offer more explanation. Have a colleague note any awkwardness so you can make corrections. Make sure that what you say matches what you show and that there are no errors (facts, formats, spelling, etc.) in what you show.

Talk to your colleagues, not over their heads or "at" them. The easiest way to do this is to pick four or five interested people in different parts of the room and talk directly to each of them for part of the presentation. Remember that you are speaking to an audience. If you were a member of that audience, you would appreciate a presentation that is clearly communicated to you in conversational language. Use a well-modulated speaking voice and a conversational tone. Practice using a microphone with someone in the room to find your best public speaking volume. Speaking too softly or too loudly conveys inexperience. Speak clearly and articulately. Speaking too quickly conveys nervousness and lack of confidence.

In most meetings, you will have a limited amount of time to present your work. Rehearsing is also a way to make sure that your presentation fits into the allotted time. If you go over your allotted time, you will waste your colleagues' time and they will resent it. In many meetings, if you go over time by more than a minute or two, the chair will ask you to stop speaking. When this happens, the only thing the audience will remember is that you had to be asked to stop. Prevent this by having someone help you time your rehearsals and then trim or expand the material until you can keep your presentation within the allotted time in front of the mirror at home or in front of an audience for rehearsal.

To prepare answers to possible questions, you can ask your rehearsal audience to think of questions that might come up, or you can discuss your research with anyone who will listen. You can use your fellow students, colleagues, friends, family, etc. to practice discussing your research at different levels. They may have useful insights, or you may find that verbalizing your ideas clarifies them for you. Practice with your visual aids. Many presenters undermine their own talks by being clumsy with visual aids and materials. Practice pointing to the image on the screen and turning back to the audience. If using transparencies, practice positioning the sheet and then moving out of the way so that your body does not block the audience's view. If possible, practice in a room that is close to the size of the room you will be presenting in. Practice your talk with your slides or transparencies until you can practically ignore your notes. Always keep a mental picture of the audience, imagine you are talking to them. More rehearsals will give you the opportunity to experiment with different techniques in front of a live audience without losing your professional credibility.

If possible, videotaping a practice session is a guaranteed way to polish your presentation. Run the videotape over and over again to look for these things: Are you making eye contact? What are your hands doing? Are you smiling occasionally? How is your posture? Do you notice any distracting mannerisms? Are your thoughts logical? Are your transitions smooth? Do you vary your voice and pace for emphasis, to avoid monotony, and as you transition to new thoughts? Do you hear any "errs" "ahs", or "ums"? Are you using the pointer correctly? This method of commenting on and improving your presentation is extremely valuable. Remember, you can never get enough practice.

Rehearse on site if possible. Schedule a technical and final dress rehearsal with the audio/visual crew. This will help you familiarize yourself with the location and equipment and avoid any problems that may arise during the formal presentation. Do a last-minute check of your manuscript, disk, slides, and any other special needs, such as a pointer.

The most effective speakers exude charisma and relaxed confidence. To be most effective, the scientific speaker must develop a delivery style that includes good body language, pleasant facial expressions, and a confident and positive yet relaxed tone of voice[4]. These skills cannot be "taught" or imposed from the outside. However, under certain conditions, these qualities can be "captured" and developed. The more speaking techniques you learn and practice, and the more you read about building a mature mind, the sooner you will build on your natural style and develop your full potential as a public speaker. The suggestions in this section and in the sections below can be very helpful in preparing an effective presentation. As you rehearse, however, keep in mind that you are not putting on a dramatic show like an actor or politician. Stage fright develops from a performance-oriented approach. When you shift from performance-oriented speaking to relationship-oriented speaking, speaking one-on-one, friendly, personal, real, conversational with each person in the audience, you will find more of yourself and become more comfortable, confident, authentic, and powerfully positive with your audience, allowing you to deliver your message in the most effective way possible.

② Culture considerations and manners. No matter what cultural and academic environment you are in, good manners, respect, courtesy, and consideration for others can make communication much easier. However, in a cross-cultural setting, you will sometimes need to be aware of different cultural considerations. For example, in India, putting your arm around a person of the same sex on the street is nothing out of the ordinary, while public displays of affection to the opposite sex are unacceptable to the local people. Another example is that people from most cultures outside the United States are more interested in discussing personal matters such as family, income, housing, and other non-business matters during conference breaks than most Americans. Many values and behaviors differ from culture to culture. Therefore, a clear understanding of different cultures is necessary to ensure better performance in cross-cultural situations.

As a researcher, there are many benefits to attending international conferences other than professional development. Do not go to a conference with blinders on, focusing only on your own career path. Do not be afraid to communicate with other participants from different countries and cultures. Cultural diversity is the spice of life, even if you are only interested in the topics within your own area of expertise. Keep your eyes, ears, and mind open. There is a lot to learn in and out of these meeting rooms, and a lot to share with your colleagues. It would be good if you could learn about our own Chinese cultural heritage and try to express Chinese thoughts, traditions, and current events in English. Otherwise, you will feel terribly awkward and frustrated when you are approached for special interests related to China if you do not know how to talk about our culture, our people and our nation in English. It is something very pleasant to share our culture with your peers from other civilizations and let them know more about China. At the same time, it can contribute a lot to your successful communication in an intercultural event.

One last piece of advice: dress for success. Do not be under- or over-dress. Dress appropriately. Choose clothes that make you feel comfortable and professional. Scientific meetings and gatherings generally allow more latitude in what is considered appropriate attire than bankers' or lawyers' conferences. Extremes, either too casual or too over-the-top, are not good ideas if you have a speaking role at a meeting.

(2) Miscellaneous information for attendance

Before attending a meeting, you need to get as much information as possible about the organizer, the host city, transportation, details of the meeting arrangements, such as working languages, whether simultaneous translation will be provided, etc.

Below you will find the "General Information" published by the Organizing Committee of the "AVS 49th International Symposium" held in Denver, Colorado, USA. From this list, you will learn many aspects that should be taken into consideration when preparing to attend a meeting.

Symposium Registration Procedures

We strongly encourage you to pre-register in advance using our on-line registration program, which will ensure faster confirmation. Those of you wishing to register by mail or fax may download the form and mail it to the following address to be received no later than October 14, 2002.

Please note that all domestic pre-registrants will receive their badges and event tickets by mail in advance of the meeting. You must report to the Badge Holder Pick-up Counter on-site to collect your holder.

All international registrants will be required to report to the International Badge Pick-up Counter in the registration area to collect their meeting materials.

Symposium Registration Cancellation Policy

All symposium cancellation/refund requests must be submitted by Wednesday, October 30, 2002, in writing to …

Please note that all refunds will be issued via check within 30 days after the meeting. All refunds are subject to a $25 cancellation fee.

Symposium Lost Badge Policy

Please note that we will be imposing a $20 fee for replacement badges, so please remember to bring your badge and keep it in a safe place throughout the week.

IMPORTANT NOTES

Please note that AVS will not publish an Abstract Book this year, so you must refer to the website to review and print abstracts of interest. All abstracts as well as a personal scheduler have been available on our website since early July.

Computer terminals with a program link will be available in the symposium registration area should you need to reference any abstracts during the week.

Election Day for U.S. Citizens is Tuesday, November 5th. Voters should secure their absentee ballots prior to leaving for the Symposium.

Hotel Reservations

To make hotel reservations, you may register on-line or you may complete and return the official HOUSING FORM to the HOUSING BUREAU by October 3, 2002.

*ADAM'S MARK (SYMPOSIUM HEADQUARTERS)—located at 1550 Court Place and just three blocks from the Olorado Convention Center. The AVS rate is $127 for single; $146 for double; $165 for triple; and $164 for quad.

Ground Transportation Information

The Denver International Airport is 28 miles (approximately 40 minutes, 1 mile = 1609.344 m) from downtown Denver. A variety of transportation services to and from the airport are available. You may contact the Ground Transportation information counter at 1-800-AIR-2DEN or locally at 303/342-4059. Options include public transit, scheduled door-to-door shuttle van service, charter buses, limousines, rental cars, taxicabs, hotel courtesy shuttles and wheelchair services.

Travelers may also obtain ground transportation information at the Ground Transportation Information Counter located on Level 5 of Jeppesen Terminal.

Shuttle Service Information

Denver provides FREE shuttle service on the 16th Street Mall, pedestrian street adjacent to the Adam's Mark Hotel. The shuttle runs from 5:00 a.m. to 1:00 a.m. Monday through Friday. (It begins at 7:00 a.m. on Sunday and 5:30 a.m. on Saturday.) The time between shuttles is 3-4 minutes (although this time is somewhat longer after 9:00 p.m.). The shuttle DOES NOT RUN DIRECTLY to the Convention Center. However, you can take the shuttle from the Adam's Mark Hotel to Welton or California Street — a saving about 4 blocks of this journey to the Convention Center. The shuttle runs from Civic Center (Bus) Station (Colfax and Broadway) near the Adam's Mark Hotel to Union (Train) Station (17th and Wewatta Streets).

Complimentary Membership Offer

If you have paid the $550 or $590 nonmember registration fee, you will automatically receive a complimentary AVS electronic membership for 2003. You do not need to fill out a membership form as your symposium registration contains the information that is needed to begin a membership in AVS. After the symposium you will be contacted by email asking in which divisions you would like to participate. If you wish to receive a *JVST* print subscription you must pay any additional fees. For more information, stop by the AVS Education Center during the symposium or contact…

JVST Manuscript Publication

The papers from this Symposium will be published in the May/June 2003 issue of the Journal of *Vacuum Science and Technology*. Active 2003 members will receive a copy of the Journal. Accepted authors interested in publishing, please contact ...

Recording Equipment Policy

Use of video recording equipment, cameras, or audio equipment at any AVS International Symposium, Short Course, or Topical Conference is prohibited without prior written approval of the AVS. The AVS reserves the right to reproduce, by any means selected, any or all of these presentations and materials.

22nd Annual AVS Run

The 22nd Annual AVS Run will be held on Wednesday, November 6th at 6:30 a.m. A map identifying the starting point will be provided at the race registration desk. The race is a 5 km course. Applications for the race are available online or may be obtained from the race director ...

The run fee is $20.00 US (late fee of $22 after October 28). This includes a high-quality, commemorative tee-shirt, race numbers, awards (in many age categories) and post-race refreshments. All runners, competitive or a bit less serious, are invited to participate.

The CORPORATE DIVISION RACE will be held again. Each team needs at least three members and times are handicapped by age and sex. Teams must be designated in advance of the run. Non-AVS members will be time-penalized in this corporate event, so make sure your membership is current! Contact the race director to register your teams.

Climate Information

Nothing about Denver is more misunderstood than the city's climate. Located just east of a high mountain barrier and a long distance from any source of moisture, Denver has a mild, dry and arid climate. The city receives only 8-15 inches (20.3-38 cm) of precipitation per year (about the same as Los Angeles), and records 300 days of sunshine per year — more annual sunshine hours than San Diego or Miami Beach. Fall weather in Denver has warm days and cool evenings—although late fall can bring some snow. Daytime temperatures can range from warm to brisk, so several layers of moderate-weight clothing will allow adjustment for maximum comfort. Drinking plenty of water will help offset the effects of altitude, especially altitude headaches. Drinking alcohol, on the other hand, should be done in moderation as its effects are enhanced at Denver's 1-mile high altitude.

AVS Message Center — new service!

The Message Center is a service for you to communicate with your colleagues while attending the AVS 49th International Symposium. This feature allows you to search the registration list for the person you want to communicate with, and once you find the person in the registration list, you can send them a message. The Message Center also allows you to read messages that people have sent to you, with the ability to reply. If you see your name scrolling on the Message Center kiosk, then you have a message and you can choose to read the message then, or you can access or send messages remotely from the comfort of your hotel room while attending the event.

This gives you a quick and easy way to network and schedule meetings during the week.
Look for the message center kiosk in the registration area.

2003 Symposium Proposals

We are pleased to welcome you as a participant in the 49th International Symposium. We hope that you will be satisfied with the content of the meeting, whether your primary interest is in the technical sessions with over 1,300 papers, the short courses, the equipment exhibition, the social program, or the special events and functions scheduled throughout the week. The AVS officers and directors, through the Program and Local Arrangements Committees, strive to ensure that the International Symposium meets the needs of the community using all aspects of vacuum science and technology. To assist them in achieving this goal, suggestions for improvement in any aspect should be directed to …and… who are the Program and Local Arrangements Committee Chairpersons, respectively, for the **AVS 50th INTERNATIONAL SYMPOSIUM to be held at the BALTIMORE CONVENTION CENTER BALTIMORE, MD, NOVEMBER 3-7, 2003.**

New Words and Expressions

intuitively	[ɪnˈtjuːɪtɪvli]	*adv.* 凭直觉地
channel	[ˈtʃæn(ə)l]	*n.* 频道，电视频道，管道，渠道，手段
plagiarism	[ˈpleɪdʒərɪzəm]	*n.* 剽窃，抄袭，剽窃作品
annotation	[ˌænəˈteɪʃ(ə)n]	*n.* 注释，评注，加注
bibliography	[ˌbɪbliˈɒgrəfi]	*n.* 书目，参考文献
colloquium	[kəˈləʊkwiəm]	*n.* 座谈会，学术研讨会
convention	[kənˈvenʃ(ə)n]	*n.* 习俗，大会，集会
convocation	[ˌkɒnvəˈkeɪʃn]	*n.* 召集，集会，教士会议，评议会
document	[ˈdɒkjumənt]	*n.* 文件，文献，单据
venue	[ˈvenjuː]	*n.* （事件的）发生地点，（活动的）场所
solvency	[ˈsɒlvənsi]	*n.* 偿付能力，溶解力
banquet	[ˈbæŋkwɪt]	*n.* 宴会，盛宴；*v.* 宴请，参加宴会
logistics	[ləˈdʒɪstɪks]	*n.* 后勤，组织工作，物流
solicit	[səˈlɪsɪt]	*v.* 请求，索求，征求（意见）
plenary	[ˈpliːnəri]	*adj.* 充分的，全体出席的；*n.* 全体会议
moderator	[ˈmɒdəreɪtə(r)]	*n.* 仲裁人，调解人，（会议或讨论的）主持人
panelist	[ˈpænəlɪst]	*n.* 专门小组成员，专题讨论小组参加者
prevail	[prɪˈveɪl]	*v.* 盛行，流行，普遍存在
exhibitor	[ɪgˈzɪbɪtə(r)]	*n.* 展出者，显示者
plasma	[ˈplæzmə]	*n.* 等离子体，血浆，等离子气体
oral	[ˈɔːrəl]	*adj.* 口头的；*n.* 口试
width	[wɪdθ]	*n.* 宽度，广度
bottom	[ˈbɒtəm]	*n.* 底部；*adj.* 底部的

column	[ˈkɒləm]	n. 圆柱，纪念柱，柱状物，纵队，纵行（列）
punctuation	[ˌpʌŋktʃuˈeɪʃn]	n. 标点符号，标点符号用法
summarize	[ˈsʌməraɪz]	v. 总结，概述
static	[ˈstætɪk]	adj. 静止的，静态的；n. 静电
affiliation	[əˌfɪliˈeɪʃ(ə)n]	n.（与政治、宗教等组织的）联系，从属关系
categorize	[ˈkætəgəraɪz]	v. 将……分类，把……列作
foreground	[ˈfɔːgraund]	n. 前景，最显著的位置
font	[fɒnt]	n. 字体
correctly	[kəˈrektli]	adv. 正确地，合适地，得体地
disharmony	[dɪsˈhɑːməni]	n. 不一致，不调和，不融洽
piece	[piːs]	n. 张，片，碎片，（成套物品的）部件，部分
sequentially	[sɪˈkwenʃəli]	adv. 从而，继续地，循序地
proposal	[prəˈpəʊz(ə)l]	n. 提议，建议，提案，（计划、建议等的）提出
reimbursement	[ˌriːɪmˈbɜːsmənt]	n. 报销，补偿
endorsement	[ɪnˈdɔːsmənt]	n. 背书，票据签字
certificate	[səˈtɪfɪkət]	n. 证明，证书
rehearse	[rɪˈhɜːs]	v. 预演，排演
pronunciation	[prəˌnʌnsiˈeɪʃ(ə)n]	n. 发音方法，读法，发音
mirror	[ˈmɪrə(r)]	n. 镜子，写照，真实反映
allotted	[əˈlɒtɪd]	v. 分配，指派，拨给（allot 的过去分词）；adj. 专款的，拨出的
rehearsal	[rɪˈhɜːs(ə)l]	n. 排练，排演，预演，演习，复述，重复
clarify	[ˈklærəfaɪ]	v. 澄清
clumsy	[ˈklʌmzi]	adj. 笨拙的，不得体的，难处理的，难使用的
mannerism	[ˈmænərɪzəm]	n. 言谈举止，特殊习惯
blinder	[ˈblaɪndə]	n. 眼罩
symposium	[sɪmˈpəʊziəm]	n. 专题研讨会，讨论会，专题论文集
limousine	[ˈlɪməziːn]	n. 豪华轿车，大型豪华轿车
taxicab	[ˈtæksiˌkæb]	n. 出租车
shuttle	[ˈʃʌt(ə)l]	n. 航天飞机，往返于两地的交通工具，摆渡车
handicapped	[ˈhændikæpt]	adj. 残疾的，智障的，残障人士专用的
moisture	[ˈmɔɪstʃə(r)]	n. 潮气，水分
vacuum	[ˈvækjuːm]	n. 真空，v. 用真空吸尘器打扫
professional journal		专业杂志
membership		会员身份

society	学会，协会
association	协会，学会
federation	联合会，协会
on-line	在线的，即时的
conventionally	按照惯例地
affiliation	加入，从属关系，分公司，分部，附属机构
cf. organization, institution, company	
cover sheet	封面
deadline	截止日期
e.g. be postmarked no later than	以邮戳为准不晚于
cf. Papers should be received no later than…	
Submission is due on…	
The deadline is by…	
The registrations close on…	
The deadline for registration is…	
get registered by…	
pre-abstracts, advanced abstracts	会前文摘预印集
proceedings	会议论文集
cf. symposium, papers, transactions, reports, prints	（均可表示会议论文集）
host	主办机构（如主办学校、主办城市、主办国）
plenary session	全体会议
companion/companying person	随行人员
copyright policy	版权规定
secretary	秘书
secretariat	秘书处
organizing committee	组委会
program committee	程序委员会
scientific committee	科学委员会
logistics committee	后勤委员会
session chair	会议主席
moderator	会议主持人
technical program chair	技术程序主席
keynote session	主题会议（一般为全会）
panel session	专题讨论会、分组讨论会
cf. concurrent sessions, simultaneous sessions, panels, discussions, symposia, colloquia, satellite meetings, topical meetings, thematic meetings	
research-in-progress session	
general discussion	一般讨论，综合讨论
one-on-one Q & A	一对一问答
oral presentation	口头报告

poster presentation	海报展示
first-come, first-served	按先后顺序
time slot/time limit/time block	时间段
cf. allotted time	
allocated time	
agreed time limit	
stay within…minutes	
confine to…minutes	
panelist	专门小组成员
direct photographic reproduction/camera-ready printing	照相直排印刷
uniform style	统一格式
laser printer	激光打印机
gutter	文章两栏间的空格
typeface and typesize/font and size	字形，字号
single spaced/spacing	单倍行距
1.5/double spaced/spacing	1.5 或双倍行距
indenting	缩进（行、段落、列表等）
Times New Roman 12 point	新罗马字体十二号字（这是打印英文文本的最常用字体与字号）
italics	斜体
bold	粗体
underlined	下划线
registration fee payment	注册费付款
references	参考文献
cf. references citations, references cited, works cited, bibliography	
submission/contribution	投稿
review	审稿
conference fee	会务费
registration	注册
recreational activities	娱乐活动
accommodation	食宿费
transportation	交通费
airfare	机票费
transfer	转乘、转机
shuttle	区间车（如校车、往返市区与机场之间的班车等）
tax	税费（许多国家在标价之外要收取一定比例的税费）
VAT	增值税（value added tax）
gratuity/tip	小费

incidental expense	杂费
travel expenses to and from (a city)	往返某城市的旅费
studentship	学生身份（同会员身份一样，学生身份也是可以获取某些资助的渠道之一）
National Foundation of Sciences	国家自然科学基金
National Foundation of Social Sciences	国家社会科学基金
waived or reduced registration	减免注册费
graduate/postgraduate	研究生
master student	硕士研究生
doctoral student	博士研究生
doctor candidate/Ph.D. candidate	博士候选人
study (be) in a doctoral program	攻读博士学位
study for a doctoral degree	攻读博士学位
postdoctoral researcher (fellow, associate)	博士后（简称 postdoc）
advisor/supervisor	导师
letter of endorsement	由某人签名的保证书、证明函等
advance/early registration	会前预先注册
early bird registration	提前注册
on-site registration	开会时现场注册
on-line registration	网上注册
best student paper award	最佳学生论文奖
critical professional skill	重要的专业技能
specific terms	专业术语
eye contact	（与观众之间的）目光交流
working/official language	大会工作语言
simultaneous translation	同声翻译
official housing form	正式住房表
the AVS rate	为 AVS 这次会议特定的房价
free shuttle service	免费区间车服务
each conference retains the copyright to their proceedings but allows that copies of single articles may be taken from this cite, provided that the source is acknowledged	每个会议拥有其论文集的版权，但允许使用本网站上的单篇文章，只要使用时注明出处即可
invited papers are allotted 25 minutes for presentation and 15 minutes for questions	分配给特约发言人的报告时间是 25 分钟，问答时间是 15 分钟
run in a single-track format	在一个时间段内只安排一场会议，不安排其他平行会议的开会方式
character	西文一个印刷符号，如"A""b""6""%"；中文一个汉字
all text should be justified/aligned	全文须左右均对齐
left justification, ragged on the right margin	左对齐，右不对齐
flush/place/put against the left margin	左对齐

centered	居中
justify the text	左右均对齐
aligned left and right	左右均对齐
full justification	左右均对齐
whole justification	左右均对齐
indent the first line of each paragraph one em-quad. Do not indent after a heading or an open line	每一段的第一行要缩进一个字符的宽度。在标题之后或空行之后不要缩进
in lower case	用小写字母（此要求用于标题时，标题的第一个词的第一个字母一般仍需要大写。）
initials should precede the surname	名的缩写应放在姓氏之前
leave two blank/open/white lines before starting the text of the paper	论文正文前应有两行空行
type primary headings in capital letters Roman and secondary headings in lower case italics	大标题以 Roman 体大写字母打印，小标题以小写斜体打印
when using abbreviations, spell out in full at first mention, followed by the abbreviation in parenthesis	使用缩写时，应在第一次提及时写出全称并在括号里注明缩写
acceptable for publication with minor revisions	稍微修改即获采用
unacceptable /rejected	退稿
culture considerations	文化因素
c/o	= care of 请转交（c/o 前面是以此地址收信的人，其后是该信的收信人）
you must report to the Badge Holder Pick-up counter on-site to collect your holder	您必须到现场的徽章领取柜台领取您的徽章
please note that all refunds will be issued via check within 30 days after the meeting	请注意，所有退款均在会后 30 天内以支票方式发放
all refunds are subject to a $25 cancellation fee	所有退款均须扣除 25 美元注册取消费
election Day for U.S. citizens is Tuesday, November 5th. Voters should secure their absentee ballots prior to leaving for the symposium	美国公民参加选举的日期是 11 月 5 日（星期二）选民要在前来开会前安排好自己缺席选举人票投票事宜
AIR-2DEN	这是一种记忆电话号码的标识法，此号码按其读音为"Air to Den(ver)"，拨打电话时按电话上相应的字母键即可，AIR-2DEN 即为 247-2336。又如，电话号码为 8294-4968 的出租汽车公司可用"TAXI-4YOU"；号码为 866-9233 的名为 Thomas Wade 的人，也许会说："Please call me at TOM-WADE"
if you have paid the nonmember registration fee, you will automatically receive a complimentary membership for 2020	已交付本次会议非会员注册费者，将自动获得协会赠予的 2020 年度会员资格
…is prohibited without prior written approval	如无事先书面批准，则不允许……

handicapped by age and sex　　　　　　　　根据性别和年龄，给优者不利条件、给劣者有利条件，以使得胜机会均等的竞赛

Notes

1. You will be well informed about the latest developments reported at past professional meetings in your field if you pay close attention to the conference literature printed during the meeting, or printed before or after the meeting.
   如果你密切关注会议期间印刷的会议文献，或者在会议前后印刷的文献，你将会很好地了解在你的领域已经举行的专业会议上报告的最新进展。

2. When pasting the figures, remember to place the actual lines of the figure immediately against the top of the type area: ignore the open space which may be present above the lines of the figure.
   粘贴图形时，请记住将图形的实际线条紧贴版面的顶部；忽略图形线条上方可能存在的空白区域。

3. The stipend will be paid either in advance of the conference or as a reimbursement, in which case you will pay for travel expenses and then submit a voucher and original hotel and transportation receipts documenting those expenses.
   津贴可在会议前支付或会后报销，当会后报销时，您需要先支付差旅费，然后要提交一份住宿和交通费用的支付凭证进行报销。

4. To be most effective, the scientific speaker must develop a delivery style that includes good body language, pleasant facial expressions, and a confident and positive yet relaxed tone of voice.
   为达到最佳效果，科学演讲者必须形成一种演讲风格，包括良好的肢体语言、愉悦的面部表情以及自信、积极但轻松的语调。

Trivial — Last but not the Least

Postdoctoral Fellowships for International Scholars

国外博士后奖学金分为：1. 外方导师资助；2. 中国政府公派，国家留学基金委（CSC）、博管办博士后资助项目；3. 国外机构的博士后奖学金项目，如德国洪堡、欧洲玛丽居里、日本 JSPS 等。外方导师资助是最优的选项，其中，如何有效地与外方导师联系的相关英语知识已经在本单元有很多介绍；中国政府公派、CSC、博管办资助的项目都可以在中文网页上找到信息，在此不赘述。现在用最简单的方式介绍一下第三类，国外机构的三个博士后奖学金项目，希望对有申请博士后研究需求的同学有所帮助。这些信息均来自官方网络。

名称	年录取人数	申请条件	申请专业	资助额度	研究期限	申请日期
洪堡	约 500	博士毕业 4 年内	化学/物理学/环境等	每月约 3000 欧元	6~24 月	每年 3/7/11 月前
玛丽居里	约 1000	博士毕业 5 年内		每月最高约 6000 欧元		每年 4 月
JSPS	约 200	博士毕业 6 年内		每月约 40000 日元	12~24 月	每年 5/9 月第一周

以上信息最终以当年公布的官方信息为准，具体可以参考奖学金官网。

Unit 11
Letters for Academic Communication

Letters, including emails, play an indispensable role in one's professional development. As a chemist and/or a chemical engineer, you will often need to write to various organizations, editors, or your peers for information, collaboration, positions, publishing papers, or exchanging ideas. The ability to write letters in English can be as essential to you as the ability to do research work. Good letter writing, which is an important academic asset, thus becomes a factor that will account for your success in international academic circles through writing[1]. Sample letters illustrating basic techniques and useful expressions are provided for good letter writing with different academic purposes.

11.1 Introduction to Letter Writing

The primary functions of letters are to provide a communication between two people or organizations and to preserve a permanent record of the communication. This section discusses the basic requirements for conventional letter writing and provides sample letters with common formats.

11.1.1 Basic Principles for Letter Writing

Good organization, correct grammar, punctuation, and format are the basics of good English letters. In Secretarial Duties, John Harrison provides the following guidelines for writing a good letter.

(1) Plan your message carefully so that you have a clear idea of what you want to say or ask.

(2) Write clearly and concisely, but be careful not to sacrifice courtesy of tone and precision of meaning to brevity.

(3) Write simply and directly, with just the right amount of emphasis to enable the reader to grasp the content immediately. There should be no room for ambiguity.

(4) Check the letter carefully for spelling, grammar, and punctuation errors. If there is any doubt about the spelling of a word, consult a dictionary or use a computer spell checker.

(5) Divide the letter into paragraphs, each of which deals with only one main point and is arranged as follows:

✧ An introductory paragraph to introduce the subject;

✧ The body of the letter (further subdivided into paragraphs if necessary);
✧ Concluding paragraph with a closing statement;
✧ Arrange your points so that they follow logically in their proper order.
(6) Present the letter in a clear and consistent format.

11.1.2 Styles of Punctuation and Format

Over the past century, the punctuation of the address sections of a modern business letter has been simplified. Closed punctuation, in which the ends of all lines in the salutation and inside address have either a comma or a period, is obsolete in the United States and some other English-speaking countries. It has been replaced by mixed punctuation, which requires a colon or comma after the salutation and a comma after the complimentary close, but nothing at the line ends in the addresses. Another, more contemporary choice is open punctuation. In this style, all punctuation is omitted at the end of each line of the return and inside the addresses, the salutation, and the complimentary close. To more conservative writers, the absence of a colon or comma after the salutation and the omission of the comma after the complimentary close look like mistakes. The choice depends on the individual or office style; either is correct.

Three styles of letter format are currently used for formal or business English letters. They are: full block style, half block style, and half indented style.

(1) Full block style

The full block style is popular because it saves typing time. Because all lines in this format are blocked or begin at the left margin, no counting or centering is required, and there are no margin changes. Some writers object to the left-leaning, rather unbalanced appearance of this style.

Open punctuation is often used in the full block style; mixed punctuation is also acceptable. In this style, all elements (headings, date lines, internal addresses, etc.) and paragraphs are separated by at least one open line.

(2) Semi-block style

The semi-block style contains three elements that are placed in different positions than their counterparts in the full block style. These are:

✧ The date (or date and return address if there is no letterhead), which is either centered below the letterhead or placed slightly to the right of center.

✧ The closing and signature lines, which are centered or slightly to the right of center.

Either mixed or open punctuation is appropriate in this style, and a colon is more often used after the salutation. For some correspondents, this style presents a more balanced appearance than the full block form.

(3) Semi-indented style

The semi-indented style is a bit more traditional than the two styles previously discussed because it requires paragraph indentation, usually four or five spaces. With this style, paragraphs are either separated by double lines spacing or not. Mixed punctuation is usually used with this style, and a comma is more often used after the salutation.

11.1.3 Necessary Parts of a Formal Letter

All formal letters must include the items listed below.

(1) Letter head or return address

Most organizations create a printed letterhead that includes their address and communicates their image. If no letterhead is available, type your return address. It must be complete so that the reader can reply if he or she wishes. It can be placed in the upper left-hand corner (full block) or upper right-hand corner (half indented and half block), defining both the top and side margins. Your name is not necessary here because it will appear in the signature section. Study the following example:

- ◆ School of Chemistry and Chemical Engineering (branch)
- ◆ Kunming University (organization)
- ◆ 2 Puxin Road (number and street)
- ◆ Kunming 650214 (city and post code)
- ◆ P.R. China (country)

Note that there is no comma or period at the end of each line, and no "No." before the street number. If it is a well-known city, such as Nanjing and Guangzhou, the province name can be omitted. You can include your phone number, and email address if you expect a quick response to your letter. Study the following example:

- ◆ Energy Frontier Research Center
- ◆ Department of Civil and Environmental Engineering and Earth Sciences
- ◆ University of Notre Dame
- ◆ Notre Dame, IN 46556 (city, state and zip)
- ◆ Telephone: 1-574-631-7852 (telephone number)
- ◆ Email: pcburns@nd.edu (email address)

Note that if there are more than two items on a line, such as city and state, there should be a comma between them, but no comma between the state/province and the zip/postal code.

(2) Date line

The date line records the date of the letter. If a typed return address is used, the date line appears immediately below it. On letterhead paper, the date line is placed at least one space below the last line of print, on the left or right, depending on the letter format. Some writers center the date below the letterhead design.

The usual order is "month-day-comma-year" or "day-month-year" without a comma. Additives such as "st" "nd" "rd" "th" are rarely used. The following are examples of the two different styles:

May 12, 2021

26 October 2021

Note that specifying the date in numbers (e.g., 1/7/2021) is in bad taste, as it can easily cause confusion, since in the United Kingdom the date in the example would mean July 1, while in the United States and some other countries it would mean the 7th day of January.

(3) Inside address

The inside address, which must match the one on the envelope, provides all necessary information about the recipient of the letter to ensure correct delivery. It is placed two or more spaces below the date line and is blocked, starting from the left margin in all formats. It contains a minimum of three lines:

① Name

② Number and Street

③ City, State, and Zip Code

Additional lines are added, usually after the first one, if the information is too extensive to fit into three sections. If a recipient has no special title, a prefix such as Mr., Miss, Mrs., or Ms. is used.

If two people are being addressed, write their names on two separate lines:

Miss Sally Smith

Mr. Robert Johnson

Be sure to spell names correctly. Do not use both a title and a degree if they are the same. Use either "Dr. Thomas Rogers" or "Thomas Rogers, Ph.D." (note the comma before Ph.D.), but not "Dr. Thomas Rogers, Ph.D". If a name is followed on the same line by a title, separate them with a comma:

Professor Sylvester Sims, Vice Chairman

or:

Professor Sylvester Sims

Vice Chairman

If you are writing to a dean, secretary, or other person whose name you do not know, you can simply write his or her job title in the inner address and then write the salutation in "Dear Sir or Madam". Study these examples:

Dr. Philip Somersby, Director

Department of Foreign Languages and Literatures

206 South Park Street

Fort Collins, CO 80523-1774

U.S.A.

The Secretary

Bureau for Educational Visits

Seymour Mews House

Seymour Mews

London W1H9PE

England

(4) Salutation

Write the salutation on the left margin, two spaces below the inside address. In almost all formal English letters, the first word of the salutation is "Dear". If the letter is addressed to an individual, use Dear Ms. X, Dear Mrs. X, or Dear Miss X for females and Dear Mr. X for males. If appropriate, use a title such as Dr. or Professor. Note that you never include the person's initials in a salutation, so do not write "Dear Mr. T. Smith".

If the letter is addressed to a job title without a specific person's name (such as Director of Personnel), use "Dear Sir" or "Dear Madam". If you do not know the gender of the recipient, you can use "Dear Sir or Madam" or "To whom it may concern". If the letter is addressed to a group, such as a company or department, use "Gentlemen or Ladies". It is followed by a colon or a comma in mixed punctuation; no punctuation follows when open punctuation is used. Study the following examples:

Dear Professor Bacon

Dear Sir or Madam

Gentlemen

To Whom It May Concern

Note that there is no "Dear" before "Gentlemen" "Ladies", and "Gentlemen and Ladies" when writing to a group.

"Gentlemen" is correct when writing to an all-male organization. "Ladies" is appropriate when writing to an all-female group. The feminist movement denounces the use of "Gentlemen" when addressing a mixed group; however, it is still preferred by some who find "Gentlemen and Ladies" (or "Ladies and Gentlemen") wordy and awkward. Dear "Sirs" is considered obsolete and is rarely used today.

(5) Body

The body of the letter contains the message. The first paragraph begins one or two spaces below the salutation or subject line, depending on the format used. In Full Block and Semi-Block styles, paragraphs begin flush with the left margin. The first line is indented by four or five spaces in the semi-indented style, or by a single space within paragraphs in all formats, or by a double space between paragraphs in some formats.

(6) Complimentary close

Place the complimentary close one or two lines below the end of the body, starting at or five spaces to the right of center in half-indented and half-block formats, or at the left margin in full block format.

The first word is capitalized. If a comma or colon follows the salutation (mixed punctuation), a comma follows the complimentary close; otherwise, both are without end punctuation (open punctuation). Here are some of the most common choices:

Sincerely

Sincerely yours

Yours sincerely

"Respectfully yours" can be used when addressing someone much older or of high rank in a government or organization.

(7) Signature

The signature section consists of a pen-written signature, a typed name, and several optional additions.

Type your name at least two spaces directly below the complimentary close. This part cannot be omitted, as it is important for the recipient to recognize your name. Many letters have gone unanswered because the writers' names were illegible. When typing this part, women may write Mrs., Miss, or Ms. in parentheses before their names to indicate their preference. Titles such as Dr. or Professor may precede the typed name. Degrees such as M.D. and Ph.D. follow the name.

Sign your name in the space between the complimentary close and the typed name. The handwritten signature should be the correspondent's full first name as he or she usually chooses to be known. Titles such as Professor and Dr., or Mrs., Mr., Ms., and Miss are not used before the signature, nor does the signer add degrees or titles after the name.

Below the typed signature, a title and/or the name of the institution (especially if it is not clearly indicated in the letterhead or return address) may be typed in one or more single-spaced, separate lines. See the following examples:

Sincerely yours
Terry Bradley
Terry Bradley, Ph.D.

Truly yours,
Weng Zhehui
(Ms.) Weng Zhehui, Ph.D.
School of Chemistry and Chemical Engineering
Kunming University

11.1.4 Optional Parts of a Letter

Certain optional items are sometimes required in addition to the required items introduced above.

(1) Reference number

Many writers of formal letters would place a reference number at the very beginning of a letter, usually in the top left or top right corner for either the block or indented style. Some stationery letterheads provide space for reference numbers. Reference numbers and letters allow responses to be lined up with previous correspondence and ensure that they reach the right person for the right file without delay. Below are some examples of reference numbers:

Reference Number: TH96008
Ref. 5678
Ref # 02-2099

You must quote the reference number in your reply letter if you find it in the letter you receive. Failure to do so may result in inconvenience.

(2) Attention line

Although not an essential element of a letter, the attention line is often used when the writer is unsure of whom to contact within an organization. It is also useful when contacting a person who has handled a particular matter in the past, especially when the writer is unsure if that person is still the appropriate addressee. The attention line can also direct a letter to the correct department in a large company, saving delivery time.

The attention line is placed two lines below the inner address and two lines above the salutation. It can be placed on the left margin or in the center of the page. All capital letters or upper and lower case may be used. The word "Attention" is followed by the title of a position or the name of a person who is responsible for the matters in the letter, for example:

Attention: Sales Manager
ATTENTION: CREDIT MANAGER
Attention: Editor

(3) Subject line

The subject line, an optional element, can help get a letter off to a fast start by eliminating the need for an introductory sentence explaining what the letter is about. When used, it is placed two spaces below the salutation and two spaces above the body of the letter. It may be placed on the left margin or centered, in all capitals, or with only the first letters of the words capitalized or underlined. The words "Subject" or "Re" are usually followed by a colon. The subject line should be short and very specific.

Dear President Johnson,
SUBJECT, DEVELOPMENT FUND

Dear Ms. Gill
(an open line)
Re: "Beyond Deconstruction" by Lin Wende, et al.

To whom it may concern,
Sixth World Conference on Women

Dear Sir or Madam:
Application for Postdoc Position

(4) Identifying initials

The dictator's and typist's initials follow the signature section. Placed two lines below the signature section at the left margin in all styles. They are especially helpful in a large office where the transcriber can be quickly located if a question or problem arises. Sometimes these appear only on the file copy. When both sets of initials are used, the transcriber's initials usually follow the dictator. If only one set is used, it usually indicates the typist. Acceptable arrangements include:

TK, emh

TH/emh

emh

(5) Enclosure notice

When additional items are enclosed with a letter, an enclosure notice must appear in the left margin two spaces below the identifying initials. The words "Enclosure" "Enc.", or "Encl." may be used when a single item is enclosed; "Enclosures" "Encl. 2", or "Enc. 5" indicates that two or more items are submitted.

Enclosure

Encl.

Enc.

Enclosures

Some writers list the additional contents:

Enclosure: A Copy of the Abstract

Enclosures: 1. Resume

2. Application Form

(6) Carbon-copy notation

If someone other than the addressee is to receive a copy of the letter, write a carbon copy notation. Place it on the left margin, two spaces below the previous notation. Use the following form:

cc: Mr. Anthony Canteras

(7) Postscript

Postscripts generally do not belong in formal correspondence because they indicate disorganization and carelessness. A properly planned letter should not need to add information in this way, suggesting an afterthought. However, there are occasions when a postscript is intentionally included to emphasize or reemphasize an idea that may have been overlooked in the body of the letter. Use this element wisely, well, and sparingly. It should be the last item to appear on a letter and is placed on the left margin. The initials "PS" may or may not appear.

PS: Remember, you must return the enclosed card by May 15th to take advantage of this promotion.

P.S. Do not forget to fill in the enclosed card! July 8th is the deadline. Please come for me at 8:00 pm on May I at the Lukou Airport.

(8) Continuation page headings

When a letter is very long, it is necessary to use two or more pages. These pages are usually typed on non-letterhead stationery of the same quality as the first sheet. The left and right margins are retained; the heading of the continuation page begins one inch from the top, two spaces above the first line of the continuation letter. The heading should include the addressee's name, page number, and date in one of the following styles.

Professor Edward Synge
Page 2
June 6, 2021

Professor Edward Synge 2 June 6, 2022

Never carry less than three lines over to a continuation page, and leave at least two lines of a new paragraph at the bottom of a previous page. If you handwrite, make sure you leave at least one inch at the bottom. Do not hyphenate the last word at the bottom of a page. Some of these limitations may require reformatting or reordering.

11.1.5 Stationery and Envelope

For business letters, letterhead stationery is preferred, especially when necessary. For informal letters, you can use plain paper or even colored paper with flowers or cartoons as a background. No matter what writing paper you use, letterhead stationery, personal stationery or plain paper, make sure that it is of good quality, which is a way to show respect to the reader of the letter. The layouts on printed stationery, as shown in the sample letters in the next section, are preferable to most formal letters.

On the envelope, it is important to place the recipient's address so that it can be read by the post office's optical scanners. The block of writing or typing should lean toward the right side of the envelope, slightly below the center, with at least a little white space to the right and bottom. The return address, if not pre-printed, should be typed or written in the upper left-hand corner, leaving a one-inch margin on the sides. The outside address must match the inside address of the letter.

11.1.6 Sample letter with Common Formats

UNIVERSITY OF NOTRE DAME
College of Engineering
CIVIL & ENVIRONMENTAL ENGINEERING & EARTH SCIENCES

To Whom It May Concern:

This letter is to verify that Dr. Zhehui Weng is employed at the University of Notre Dame as a Postdoctoral Research Associate in the Department of Civil & Environmental Engineering & Earth Sciences. Her appointment has been scheduled from February 16th, 2011 to July 31st, 2014.

She has participated in the research projects of the Energy Frontier Research Center in University of Notre Dame. She is a valued member of our group. She is now planning on returning to China for future work in April, 2014.

If you need further assistance feel free to contact me.

Sincerely,

Peter C. Burns

Director, Energy Frontier Research Center
Henry Massman Professor
Department of Civil and Environmental Engineering and Earth Sciences
Concurrent Professor, Chemistry and Biochemistry
University of Notre Dame
Notre Dame, IN 46556
Tel. (574)-631-7852
Email: pburns@nd.edu
petercburns.com
msa-efrc.com

156 FITZPATRICK HALL OF ENGINEERING
NOTRE DAME, INDIANA 46556 USA
TELEPHONE (574) 631-5380
EMAIL CEEES@ND.EDU — WEBSITE CEEES.ND.EDU

11.2 Introduction to Email Writing

Traditional formats and styles for writing letters are very important, as we discussed in the previous unit. However, in today's electronic age, email, the lesser form of communication, has become more and more indispensable in the lives of professionals. Many websites offer free email services. No printing and no postage make it almost free to send emails around the world. And minutes after you send your messages, you can start seeing responses. In fact, no one would bother writing on paper except for formal documents that require a handwritten signature.

Email, the electronic letter, is a kind of "instant communication" that allows you to make yourself understood to others in a very short time. It is more conversational than the traditional letter. Because of this, it is not given the same amount of care and attention as putting pen to paper to write a traditional letter. However, if you use email well, it can contribute greatly to the success of your career. In this unit, you will find some guidelines for writing an effective email.

11.2.1 Email Format

Not all computer screens and email systems are created equal. The format in which you type your email may be changed when it reaches its destination. Therefore, the recommended format for typing email is the full block style. Flush everything, including the salutation, the body, and the sender's name, to the left, leaving open space below the salutation, above the sender's name, and between different paragraphs[2]. The sender's email address and the date are not necessary, as they are provided automatically. Other elements must be typed into the columns provided by the email system.

11.2.2 Subject Heading

The subject line should be short and to the point. It is one of the first things your reader will see. It is something that will distinguish your message from many others in the mailbox. Make it count. A subject line like "Hi" or "Hello" is just as useless as nothing at all. Instead, something like "Hello from Kunming" or "Hello from Wang Ping" or, more specifically, " Wang Ping's Travel Plan" will make more sense and attract more attention.

When there are back-and-forth messages on the same subject, it will be easier to keep track of things if you change the subject line according to the current stage of the conversation. For example, if your first message is about submitting a paper, you might use "Submission from Liu Ping" or "China's Environment by Liu Ping". For the second message, do not use the same subject line. Instead, you can use "Liu Ping's Resubmission" "China's Environment Revised", or "MS-AB123 Revised". For further correspondence, you can subject your messages as "Second Revision of MS-123" "Final Version of MS-123" "Thank You from Liu Ping", and so on. This way, just by reading the subject line, the recipient can get an idea of the main point of your message or where the conversation is going.

11.2.3 Salutation

As in regular letter writing, the salutation, when used well, is an effective way to set the tone of the letter. Think of it as a handshake, another way of greeting the recipient. It puts the reader in the right frame of mind and prepares him for your message.

If you normally address a person as Prof./Dr./Ms./Mr. Smith in a regular letter, then that is how you address them in an email. If you normally call them by their first name, you can do so or simply omit the salutation by saying "Hello" or "Hi, Mary", or if you are very close friends or fellow students, you can use a more casual "Hi there". However, if you are unsure, especially if you are writing an email for formal correspondence, it is safe to stick to the formal salutation.

11.2.4 The Body

In a regular letter, it is important to make things clear with complete sentences and standard grammar. But because email is a quick way to communicate, your writing style can change depending on your audience. You can freely combine aspects of informal speech, formal written communication, and new ways of showing emotion and body language in an email message.

Typically, people find it harder to read words on a computer screen than on paper. And more importantly, you are communicating quickly, so you should make your email easy to read, with a style that is a little different from a regular letter. Here are some helpful suggestions.

(1) Use shorter paragraphs and less words

Wordy sentences and long paragraphs are not appropriate for most email, especially informal email. Sometimes you can use abbreviated sentences without fear of breaking standard grammar. For example, you might say "Glad to get your message" or "Well done!" in a message to a friend. For both formal and informal email, consider breaking up your message into just a few sentences or paragraphs. That way, the recipient can easily see new paragraphs without scrolling to find your key points. Don't write long messages. Try to keep everything in one "screen", in about 20 to 30 lines.

(2) Show emphasis with strategies

You can use **capital** letters to emphasize your ideas or to attract the recipient's attention, e.g., "THANK YOU for ..." "Of all the students, I would like to thank YOU for ..." "Please get back to me at your EARLIEST convenience". For informal mail, you can use different ways to express emotion, body language, and intonation. For example, you can use "**smileys**": "Great! :-)" "Terrible! :-(". They are great when you really want to let your reader know what you are feeling, or how you would like your reader to feel. **Creative spelling** can also be used in casual emails: "Helloooo" " Sooorry for ..." "Teeeerribly sorry for..." "Soooo happy to…". In some cases, you can **type out thoughts** and reactions, e.g., "As to the trip, well... let me think it over…eh, to me, I…eh…I don't think it's a good idea".

(3) Use abbreviations

To save keystrokes, you can use some common abbreviations, such as "BTW" for "By the way", "FYI" for "For your information", "U" for "you" or "University", "ASAP" for "As soon as possible", "HTH" for "Hope this Helps" and "TTYL" for "Talk to you later". However, it is recommended that you use abbreviations that are already common in the English language. Otherwise, you run the risk of confusing your recipient.

Always be careful with whom you exchange messages. Use these strategies sparingly, or only in **casual messages** to your close friends.

11.2.5　Previous Message and Quotes

Some people receive dozens of emails a day. To avoid misunderstandings, you should include all or at least part of the original message(s) in your reply, or at least repeat the topic somewhere in your reply, unless you are sure that the recipient of your email knows exactly what you are talking about.

When answering questions, you can include the entire question or quote the most important part of the question. The ">" in front of the text indicates to the recipient that you are quoting material from their last email message. Study the following examples:

> I am writing a paper to present at OEC 2021. Do you remember the deadline for abstracts and the name and email address of the convener?

Yes. The deadline is October 10, 2021. Prof. Samuel Baruch's email is sbaruck@harvard.edu.

> OEC 2021 deadline for abstracts
Oct. 10, 2021.
> convener's name and email
Samuel Baruck, sbaruck@harvard.edu.

>> convener's name and email
> Samuel Baruck, sbaruck@harvard.edu.
Sorry, Baruch or Baruck?

>>> convener's name and email
>> Samuel Baruck, sbaruck@harvard.edu.
> Sorry, Baruch or Baruck?
Yes, I made a mistake. It should be "Baruch".

From the examples above you can see one level quoting (>) and multilevel quoting (>>), (>>>), etc. This is very helpful in tracing back your previous communication.

11.2.6 Closing

Before you finally give your identification in an email message, you usually end the message with a "closing", just like what you write for the closing sentences and the complimentary closing in a conventional letter. The "closing" can serve several purposes:

(1) Indicating a good will (as in informal letters)
  ◇ Best wishes,
  ◇ All the best,
  ◇ Best/Kind/Warm regards,
  ◇ Sincerely,
  ◇ Sincerely yours,
  ◇ Yours,
  ◇ With respect,
  ◇ Love,
  ◇ Take care,
  ◇ Well, good luck with your applications.

(2) Expressing gratitude
  ◇ Thanks,
  ◇ Many thanks,
  ◇ Thank you very much.
Thank you very much for your kind attention.

(3) Expressing expectation
  ◇ I'd be looking forward to your reply,
  ◇ Hope to see you next summer.

(4) Inviting questions or further contact
  ◇ Let me know if you have any other questions,
  ◇ Email me at zhweng@gmail.com instead of my zhweng@nd.edu address during my visit to

France.

(5) Saying good-bye
- ✧ Bye,
- ✧ See you then,
- ✧ So long,
- ✧ Cheers.

In addition to the examples above, all of the sample closing expressions that we will be using in the rest of this section on writing letters can be used in an email message.

11.2.7 Signature

In the "Signature" section, you will need to provide your identification. On a paper document, you must sign your name with a pen in the space above your typed name. It is not possible to sign an email message. However, you can sometimes include your title, institution, mailing address, telephone and fax numbers after your name if you are writing a business letter, especially to those who have no idea who you are. Study the following example:

CHEN Heping
Associate Professor
English Department
Nanjing University
Nanjing 210093
China
Tel: 86-25-3593243 (o),3593111(h)
Fax: 86-25-3593111

If you prefer, the email system will automatically include this information at the bottom of each email message. However, in less formal email, we do not always need to add this personal information to the name. And in everyday informal and casual email, a first name is preferable to a full name at the end. In some cases, people may even use only the initial(s) of their first name/full name to identify themselves.

11.2.8 File Attachments

Never forget to attach your file if you plan to send some documents from your computer or disk. Your recipient will be confused if they receive a message saying that a file is attached, but nowhere to find it. In this case, you will have to send a follow-up message with the file attached.

To minimize download time, it is a good idea to compress any file that is larger than 100k using programs such as WinZip. If you need to send multiple files and directories, compress them all into one attached file.

11.2.9 Reread and Save before Clicking the "Send" Button

Reread your message before clicking the "Send" button to avoid sending follow-ups to change or invent something. As a scientist, you should be careful not to make your reader think that you are being careless or redundant. Sometimes the recipient may be so confused or put off by your original letter that any of your subsequent letters will not be read at all.

If, after taking all the precautions, you still accidentally send something you did not mean to, put words like "READ THIS FIRST!" or "Revised Message from Liu" in the subject line of your revised or supplemented message. This may help you catch up on those moments of haste.

Always remember to check your spelling. If the email system you are using does not have a spell check function, you can paste your message and check it in Microsoft Word or other programs.

Before you send your message, save a copy in your "Sent" folder. This will help you keep a record of your correspondence.

11.3 Examples of Letters in Different Situations

In the previous unit, we talked a lot about the importance of attending conferences and various aspects of basic academic communication. If you are interested in attending a conference and/or have other academic goals, you may want to write and receive letters to and from the conference organizer, the editors of a journal, the collaborators you are interested in, and so on. You will often be surprised at how well and quickly your requests will be answered if you use some polite expressions. In this section, we will study some useful examples of letters related to different academic situations.

11.3.1 Examples of Letters Relating to Conference

(1) Letters of inquiry and reply

<u>Sample A.</u>   Inquiry about Main Topics of a Conference

I am writing to inquire about the "International Conference on X" that will be hosted by your center in Chicago in May 2022. I am very interested in the upcoming conference and would like to attend, as I have been working in the field of X for years.

Could you please advise me who to contact or what website to visit to get detailed information about the main topics of the conference?

I would appreciate an early reply.

<u>Sample B.</u>   Inquiry about General Information

I am a Ph.D. student in the Department of Mathematics at Nanjing University, P.R. China. I have read the *Proceedings of the* 2020 *Conference on Reasoning about Knowledge* and those of the previous meetings. Since the conference is held every two years, I assume that the next conference will be held in 2022. Therefore, I am writing for general information about the upcoming conference, such as the date and location, topics, keynote speakers, and other details. I would be very grateful if you could add me to your mailing list for the 2022 conference and for much more information about each of the upcoming conferences.

An early response would be greatly appreciated.

<u>Sample C.</u>   Inquiry about the Submission of Abstracts

I learned from your embassy in Beijing that the XYZ Congress will be held at your university in April next year. As I have been doing research on X for many years, I am very interested in attending this congress.

I would be very grateful if you could let me know the deadline for submitting abstracts. If there are any special requirements regarding the abstract, please let me know.

I look forward to hearing from you soon.

<u>Sample D.</u>　　Inquiry about the Congress Materials

I have received the acceptance letter and the official invitation from the organizing committee. However, I have not found any information about the distribution of congress materials such as programs, abstracts, etc. Could you please let me know when and where I can obtain them?

Thank you very much.

(2) Letters about expenses of the conference

<u>Sample A.</u>　　About the Possibility of Financial Support for Participation

I learned from your website that the goal of the X Conference is to discuss the latest policy developments to reduce the environmental impact of industry and to bring together participants from all countries around the world. I am a Ph.D. candidate at the School of Environmental Sciences, Nanjing University, studying industrial environmental regulation in China. I take great interest in attending this conference for the opportunity to exchange views on policies in this field with the participants from all over the world.

I noticed that you would provide financial support for a speaker and another representative from China, as the organizing committee is interested in China's particular experience in developing and implementing industrial pollution control policies[3]. The paper I am preparing is exactly on this topic, as I have had five years of experience working in a government institution in charge of environmental policy-making. I will send you my abstract and a copy of my curriculum vita to the email address you provided for the paper submission. I would be very happy if you could accept my abstract and support my participation.

I look forward to your positive response.

<u>Sample B.</u>　　Request Form for Financial Assistance

We have learned from the Second Circular for the "12th World Congress on World History" that representatives under the age of 35 who are studying in a doctoral program will be financially supported for their participation, depending on the circumstances. Could you please send us two copies of the application form as soon as possible?

<u>Sample C.</u>　　Application for Travel Assistance

Thank you for your letter dated February 10, 2020. I am writing to you about the application for my travel support.

I have discussed the problem of funding my attendance at the meeting with the director of our institute. However, he told me that the institute would only cover US $800, which is estimated to be only enough for the conference fee and accommodation. Otherwise, I am writing to you to ask for the possibility of my travel support to and from Denver.

Your kind consideration in this matter will be greatly appreciated. I look forward to your favorable response.

<u>Sample D.</u>　　Scholarship Eligibility Response

Thank you for your interest in Salzburg Seminar Session 403, "From Page to Screen: Adapting Literature to Film". Please understand that we cannot determine your eligibility for a scholarship until we receive your complete application. I am enclosing an application form and a session description for your convenience. You may also apply online at http://www.salzburgseminar.org.

If you have any further questions, please do not hesitate to contact me.

<u>Sample E.</u>   Response for Financial Support and Distribution Information

Thank you for your interest in attending ACC 2020. Unfortunately, the conference organizers are not in a position to offer financial support to delegates. I would recommend that you contact your own institution or the American Chemical Society (ACS) through their website. While they are probably not in a position to offer financial support, they can probably point you in a better direction. The fee for those who have a paper accepted for the graduate student sessions is subsidized. Full details can be found on the conference website.

Good luck with your search for financial support. We look forward to seeing you at the conference.

<u>Sample F.</u>   Response about Economic Assistance Fund and Discounted Airfares

We have received your letter of March 9, 2012, inquiring about the possibility of economic assistance. We regret to say that it cannot be emphasized enough that fundraising efforts are still underway and that there is a possibility that they may never come to fruition[4]. No specific details on the success of these efforts are available at this time. The convention organizers stress that they cannot make any guarantees. If aid is provided, the most likely form of assistance will be either a reduction or waiver of registration and/or university housing fees.

Regarding airfare, special discount codes have been established for convention attendees on several domestic and international airlines, including: X, Y, Z airlines. The travel agency listed below is the official convention travel agency. They are the best source for making your travel arrangements, both because they are familiar with the details of the convention and because they have the latest information on discount fares negotiated specifically by and for the convention.

<u>Sample G.</u>   Response for Travel Assistance and Return Ticket

We have received your letter dated August 8, 2020, requesting travel support to attend the International Conference on XYZ to be held in Sydney in October 2021. According to the terms and conditions of the conference sponsors, participants in a doctoral program are privileged to apply for financial support. Your travel expenses to and from Sydney will therefore be covered by the sponsors. However, it is expected that your institution will cover accommodation, meals, local expenses, and the conference fee.

Please note that due to the large number of people traveling to Sydney, your exact preferences may not be available. Our travel agent will advise you of your travel options, inform you of their relative costs, and then require your return confirmation. If you have any questions, please contact them at the following address.

11.3.2   Letters for Academic Visits and Cooperation

Academic visits and joint research are very important in your career development, for they are an effective way for you to communicate with your peers in the same field. In this section, letters with purposes of academic visit, cooperation and letters for exchanging ideas or materials are provided.

<u>Sample A.</u>   Letter for Exchange of Research Material

We got your name and address by surfing your institute's website and understand that you specialize in X, which is exactly the category we are studying now. This is one of the new projects in our research program, so we do not have much experience in this area. Therefore, we are particularly interested in the current reference material. We would be grateful if you could send us some of your other reports or theses on this topic.

We have some papers on the same subject published in some Chinese scientific journals, and we have translated them into English. If they are of interest to you, we would be happy to send them to you in exchange.

<u>Sample B.</u>　　Letter about Common Interest and a Planned Visit

I received your letter of January 12th, and I appreciate your interest in my research. I have compiled some of my articles to send to you. Most of them are xeroxed copies, as I have no offprints left. *The Imagery of the Cosmological Buddha* is a book and I cannot supply it, but you may have the article I wrote introducing this important sculpture.

My work has focused primarily on Southwest China (Sichuan and Yunnan), but I have also written on other geographical aspects of Chinese Buddhism. I recently published an article on Chinese cave temple sculptures, and another will appear soon on a beautiful Dali monument in Kunming, Yunnan. If you wish I will send you more after you take a look at what you will get. But to my regret, I do not know any more detailed information of the websites on Buddhism sculptures. I will come back to you after I ask someone else in this field.

Please let me know which article you want to translate and in which publication it will appear, and I will send you my written permission or permission from the publisher.

I plan to visit China during the summer holidays. I will let you know the exact dates of my visit. I hope we can meet then and discuss our mutual interests.

If you are interested in the work of our Academy, I may invite you to visit here sometime next year. We could go through all the necessary formalities to facilitate your visit.

<u>Sample C.</u>　　Invitation Letter for Academic Exchange Visit

The Department of Applied Social Studies of Hong Kong Polytechnic University invites Mr. Wang Zhixing and Mr. Zhou Xueyin to visit Hong Kong Polytechnic University for academic exchange with our social work faculty. The expected dates of their visit are May 4 to May 18, 2021. Our department will be responsible for arranging the visit and program, as well as all expenses incurred during their visit to Hong Kong.

We look forward to receiving their itinerary as soon as possible.

<u>Sample D.</u>　　Invitation Letter for a Collaborative Visit

Please accept this letter as a formal invitation to visit the United States as a guest of the Institute of Environmental Sciences. The purpose of this visit will be to study the environmental conditions of some parts of America and to discuss cooperation between your center and our institute. The places you will visit are as follows:

San Francisco/Los Angeles/Detroit New Jersey/New York/Washington, D.C.

We understand that you will be able to spend approximately 20 days in the USA. Therefore, please fax us your exact dates of availability as soon as possible so that we can make the necessary arrangements (accommodation, travel plans, etc.)

All expenses incurred will be charged to our institute. If you need further assistance, please do not hesitate to call me at my office or home number.

<u>Sample E.</u>　　Letter of Invitation for Research Collaboration

It is my pleasure to invite you and your students to visit the Department of Computer and Information Science at the University of Massachusetts. This invitation is extended on the basis of the Cooperative Research and Development Agreement we have just signed for "Machining Process Developer Software Development". The main purpose of the visit will be to implement the cooperative research and facilitate the necessary training. As stated in the agreement, all international travel expenses and living expenses in the U.S. will be provided by our department.

This invitation is extended to the following personnel from Nanjing University:

Mr. Liu Benchang, Associate Professor of the Department of Computer Science and Technology

Mr. Qian Keping, Doctoral Student of Computer Science and Technology

Ms. Zhao Yuejiao, Graduate Student of Computer Science and Technology

The duration of your group's stay in Amherst, Massachusetts is from July 15, 2019 to January 31, 2020.

I look forward to meeting you and your students and wish you a fruitful research cooperation.

11.3.3 Letters for Research Positions

In addition to short-term visits for meetings or exchanges, you can apply for long-term positions to conduct research, take courses, or teach at a foreign institution as a visiting scholar, visiting professor, or postdoctoral fellow. For these positions, you will need to obtain some scholarships to sponsor your visits, or you may also ask the host professor or institution about the possibility of hosting you as a visiting scholar. Therefore, you will need to contact your host institution or potential host to begin preparing for your visit. Below are sample letters for research and/or teaching positions.

<u>Sample A.</u>   Application for a Visiting Scholar Appointment in a Laboratory

I am an associate professor at Nanjing University, P.R. China. I am writing to you about the possibility of a visiting scholar appointment in your department next fall. I would like to take some training courses and work in your laboratory during my visit. I believe that with your help I would make significant achievements in my research.

Our government will provide all my travel and living expenses.

Enclosed are a curriculum vita of my education, research, teaching experience, publications, translations, and letters of recommendation, which I hope will meet your requirements.

I wish I could find words to express the importance of a visiting scholar position in your department, for I know how much I could learn and how much my laboratory could gain from studying and working under your guidance.

Thank you very much for your consideration. I look forward to a positive response from you soon.

<u>Sample B.</u>   Request for a Visiting Researcher Position in a Laboratory

I am writing to explore the possibility of working in your laboratory for one to two years. I am enclosing a copy of my curriculum vitae, which provides information about my educational and research experience.

As you can see from my resume, I have quite a bit of work experience in your area of research. I am confident that I can contribute to your ongoing research projects. My institute has granted me an extended leave of absence, which will allow me to gain research experience abroad. My trip to the U.S. will be partially funded by the Chinese government, which will provide me with travel expenses and a small monthly stipend (approximately $900 per month) for living expenses during the first year. Funds beyond this amount, if needed, will have to come from other sources, and if I am to stay for a second year, support will also have to come from elsewhere.

Professor Zhang Ruide, the director of my institute, and Professor Cheng Wuling, the supervisor of my Ph.D. program, have agreed to write letters of recommendation for me. They may have already written to you. If not, please feel free to contact them.

I look forward to hearing from you soon.

<u>Sample C.</u>   An Email Message for a Visiting Professor Position

I would like to apply for a Visiting Professor position in the Department of Foreign Languages and Literatures at Colorado State University. I am attaching an email version of my CV. As it indicates, I have a BA in English Language and Literature and an MA in Chinese Language and Literature. I have taught for several years, mostly Chinese as a second language. I can provide several letters of reference and recommendations from my colleagues and some of my foreign students that attest to my success as a teacher of Chinese as a second language.

My main interest is teaching students from English speaking countries. With a BA in English, I am able to teach in either English or Chinese. I believe that I have been successful as a teacher because of my dedication to teaching and my sincere concern for my students. When teaching, my classes are usually lively with lots of humor and creativity. I have taught many students from many different countries. In fact, I have taught other teachers how to teach Chinese to foreigners.

I am interested in learning more about Colorado State University and the role of a visiting professor. Please let me know what additional documents and materials you need from me.

Thank you for your time and consideration.

<u>Sample D.</u>   A Doctor Candidate's Application for a Postdoctoral Position

Please find attached all the required documents. As you can read from my CV, I do not have my Ph.D. degree certificate yet, but I will be able to take it with me to Boston in September, or I can send you a copy as soon as I receive it at the end of June.

I hope this is all satisfactory for the postdoctoral position at your center. I look forward to hearing from you soon.

New Words and Expressions

punctuation	[ˌpʌŋktʃuˈeɪʃn]	n. 标点符号，标点符号用法
comma	[ˈkɒmə]	n. 逗号，间歇
omission	[əˈmɪʃ(ə)n]	n. 省略，遗漏，被省略（或排除）的人（或物）
appropriate	[əˈprəʊpriət]	adj. 合适的，相称的；v. 拨出（款项），私占，挪用
additive	[ˈædətɪv]	n. 添加剂，食物添加剂
specify	[ˈspesɪfaɪ]	v. 明确规定，具体说明，详述
dean	[diːn]	n. 院长，教务长，主任牧师
salutation	[ˌsæljuˈteɪʃ(ə)n]	n. 招呼，致意，（信函、演讲开头的）称呼语
initial	[ɪˈnɪʃl]	n. 首字母

feminist	[ˈfemənɪst]		n. 女权主义者，女权运动者；adj. 女权主义的，支持女权运动的
denounce	[dɪˈnaʊns]		v. 谴责，指责，斥责
wordy	[ˈwɜːdi]		adj. 冗长的，口头的，唠叨的，文字的
awkward	[ˈɔːkwəd]		adj. 令人尴尬的，难对付的，不方便的
transcriber	[trænsˈkraɪbə(r)]		n. 信息转换器，转录器，誊写员
enclosure	[ɪnˈkləʊʒə(r)]		n. 围场，围墙，圈占地，附件，围住，圈起
postscript	[ˈpəʊstskrɪpt]		n. 附言，又及
sparingly	[ˈspeərɪŋli]		adv. 一点点，节省地
hyphenate	[ˈhaɪfəneɪt]		v. 用连字符连接，用连字符分割（词语）
reformatting	[ˌriːˈfɔːmætɪŋ]		n.（计算机）重新格式化
stationery	[ˈsteɪʃənri]		n.（通常有配套信封的）信纸，信笺，文具
layout	[ˈleɪaʊt]		n. 布局，设计
omit	[əˈmɪt]		v. 省去，遗漏，不做，未能做
abbreviate	[əˈbriːvieɪt]		v. 缩写，省略
scroll	[skrəʊl]		v. 卷动
strategy	[ˈstrætədʒi]		n. 策略，战略
intonation	[ˌɪntəˈneɪʃ(ə)n]		n. 声调，语调，吟咏，吟诵，（演奏或唱歌中的）音准
convener	[kənˈviːnə(r)]		n.（会议）召集人
complimentary	[ˌkɒmplɪˈmentri]		adj. 赞美的，表示钦佩的，免费赠送的
document	[ˈdɒkjumənt]		n. 文件，公文，文献，（计算机）文件，文档
disk	[dɪsk]		n.（计算机的）磁盘（= disc）
follow-up	[ˈfɒləʊ ʌp]		adj. 后续的，定期复查的；n. 后续行动，追踪调查
redundant	[rɪˈdʌndənt]		adj. 多余的，累赘的
subsequent	[ˈsʌbsɪkwənt]		adj. 随后的，接着的
precaution	[prɪˈkɔːʃ(ə)n]		n. 预防措施，防备
upcoming	[ˈʌpkʌmɪŋ]		adj. 即将来临的
embassy	[ˈembəsi]		n. 大使馆工作人员，大使馆，（外交）特别使团
delegate	[ˈdelɪɡət]		n. 代表，委员会成员
headquarter	[ˌhedˈkwɔːtə(r)]		v. 将（组织的）总部设在某地，设立总部
waive	[weɪv]		v. 放弃（权利、要求等），免除（费用）
defray	[dɪˈfreɪ]		v. 支出，支付
secretariat	[ˌsekrəˈteəriət]		n. 秘书处，书记处，秘书（书记，部长等）之职
browse	[braʊz]		v. 随意翻阅，（在商店里）随便看看，浏览信息
category	[ˈkætəɡəri]		n. 种类，范畴
xerox	[ˈzɪərɒks]		v. 复印，影印［原指使用施乐（Xerox）品牌的复印机进行复印］
offprint	[ˈɒfprɪnt]		n. 抽印本
sculpture	[ˈskʌlptʃə]		n. 雕塑，雕像
formality	[fɔːˈmæləti]		n. 手续，礼节，拘谨

incur	[ɪnˈkɜːr]	v.	招致，遭受
itinerary	[aɪˈtɪnərəri]	n.	行程，旅行路线，游记，旅行日记
duration	[djuˈreɪʃn]	n.	持续，持续时间
curriculum vitae			简历，（求职用的）履历
stipend	[ˈstaɪpend]	n.	薪金，生活津贴，薪俸，助学津贴
certificate	[səˈtɪfɪkət]	n.	证明，证书，文凭，成绩合格证书，电影放映许可证

Formal/Business Letter	正式函件、事务信函
Style of Format	格式风格
Full Block Style	全齐头式
Semi-block Style	半齐头式
Semi-indented Style	半缩进式
CV	履历表
cf. resume, vita, personal statement	
travel support to and from (a city)	往返某城市的旅费资助
return ticket/round ticket	往返票
reduction or waiver of registration and/or university housing fee	减免注册费，或减免在大学校区住宿费用，或两项费用均予减免
Single	单人间
Double	双人间
Triple	三人间
Quad	四人间
order form	订单
mailing list	邮寄名单
be included in the mailing list	
be deleted from the mailing list	

Notes

1. Good letter writing, which is an important academic asset, thus becomes a factor that will account for your success in international academic circles through writing.
   因此，良好的书信写作是一项重要的学术资产，也是您通过写作在国际学术界取得成功的一个因素。
2. Flush everything, including the salutation, the body, and the sender's name, to the left, leaving open space below the salutation, above the sender's name, and between different paragraphs.
   将所有内容（包括称呼、正文和发件人姓名）都向左对齐，在称呼下方、发件人姓名上方和不同段落之间留空白。
3. I noticed that you would provide financial support for a speaker and another representative from China, as the organizing committee is interested in China's particular experience in developing and implementing industrial pollution control policies.
   我注意到贵方将为一位发言人和另一位来自中国的代表提供资金支持，因为组委会对中国在制定和实施工业污染控制政策方面的特殊经验很感兴趣。
4. We regret to say that it cannot be emphasized enough that fundraising efforts are still underway and that there is a possibility that they may never come to fruition.
我们遗憾地指出，筹款工作仍在进行中，而且有可能永远无法取得成果，这一点无论怎么强调都不过分。

Trivial — Last but not the Least

English Letter Composition and Common Sentence Patterns
英文信件构成：

	Necessary parts of the formal letter	
	Letterhead or return address	信头或回信地址
(1)	Letterhead stationary	含有写信人所在机构名称、地址等的公用信笺
	Personal stationary	印有个人姓名、所在机构、职务、地址等内容的信笺（如在信头中已有写信人的打印名，则信尾只要手写签名即可，不必再打印一遍姓名）
(2)	Date line	日期栏
	Inside address/Addressee's address	信内地址、收信人地址
	Title/prefix	称谓（如 Professor, Dr., Mr., Ms.）
	First name/given name/birth name or its initial	名或名的缩写
(3)	Middle name or middle name initial	中名或中名缩写
(4)	Salutation	称呼
(5)	Body	正文
(6)	Complimentary close	信尾套语
(7)	Signature section	署名栏（此栏包括手写签名和打印名，也可加上写信人的职务、结构和地址等内容）
	Optional parts of a letter	
(1)	Reference/Ref.	（函件）编号
	Our ref.	我方编号
	Your ref.	你方编号
(2)	Attention line	提请注意栏
(3)	Subject line	主题行
(4)	Identifying initial	辨认记号（信的口授者和打字员姓名的缩写）
(5)	Enclosure notice	附件提示
(6)	Carbon-copy notation	抄送提示
(7)	Postscript	附言，又及
(8)	Continuation page heading	余页信头

英文学术交流信件常用句式：

1. I am writing to …
   (1) On behalf of…, I would like to…
   (2) Thank you very much for your letter dated…, inviting me to…
   (3) I have received your email of…
   (4) We learned from the last issue of Analytical Letters that…
   (5) I learned from your website that…
   (6) I learned from your letter dated May 6 that…
   (7) I am an associate professor of…
   (8) I am a Ph.D. candidate studying…

   This is a typical sentence to open a letter. Study the patterns for the opening sentences and all the opening sentences in the sample letters provided in this book.

2. An early reply is greatly appreciated.
   (1) We are looking forward to…
   (2) I look forward to seeing you this November.
   (3) Thank you very much for your consideration.
   (4) I would be very much obliged if you…
   (5) Please feel free to…
   (6) For further information, please write to…
   (7) Thank you in advance for your time and effort.
   (8) Thank you.

   This is a typical sentence to close a letter. Study the patterns for the closing sentences and all the closing sentences in the sample letters provided in this book.

3. Enclosed please find…
   (1) Attached you will find…
   (2) I enclose…
   (3) …is attached.
   (4) Attached is a…
   (5) Enclosed are two…
   (6) I am sending you…
   (7) I am enclosing…
   (8) You will find in the attachment the…

   Other similar expressions are listed here.

4. I have been working in the field for years.
   (1) We have been undertaking research on XYZ for many years.
   (2) I have been working on the Project XYZ since 2020.
   (3) We have been engaged in the research for three years.
   (4) I have been interested in…for many years.
   (5) I have studied…for years.
   (6) I have been carrying on this work since I started my Ph.D. program.
   (7) I have been involved in…for a long time.
   (8) I took up…as a profession after I got my MSc. degree in 2018.

   Other similar expressions are listed here.

Appendix

A1　IUPAC Names and Symbols of Elements

Name	Pronunciation	Symbol of the element	Atomic number	Chinese
Hydrogen	[ˈhaɪdrədʒən]	H	1	氢
Helium	[ˈhiːliəm]	He	2	氦
Lithium	[ˈlɪθiəm]	Li	3	锂
Beryllium	[bəˈrɪliəm]	Be	4	铍
Boron	[ˈbɔːrɒn]	B	5	硼
Carbon	[ˈkɑːbən]	C	6	碳
Nitrogen	[ˈnaɪtrədʒən]	N	7	氮
Oxygen	[ˈɒksɪdʒən]	O	8	氧
Fluorine	[ˈflɔːriːn]	F	9	氟
Neon	[ˈniːɒn]	Ne	10	氖
Sodium	[ˈsəudiəm]	Na	11	钠
Magnesium	[mæɡˈniːziəm]	Mg	12	镁
Aluminium	[ˌæljəˈmɪniəm]	Al	13	铝
Silicon	[ˈsɪlɪkən]	Si	14	硅
Phosphorus	[ˈfɒsfərəs]	P	15	磷
Sulfur	[ˈsʌlfə(r)]	S	16	硫
Chlorine	[ˈklɔːriːn]	Cl	17	氯
Argon	[ˈɑːɡɒn]	Ar	18	氩
Potassium	[pəˈtæsiəm]	K	19	钾
Calcium	[ˈkælsiəm]	Ca	20	钙
Scandium	[ˈskændiəm]	Sc	21	钪
Titanium	[tɪˈteɪniəm]	Ti	22	钛
Vanadium	[vəˈneɪdiəm]	V	23	钒
Chromium	[ˈkrəumiəm]	Cr	24	铬
Manganese	[ˈmæŋɡəniːz]	Mn	25	锰
Iron	[ˈaɪən]	Fe	26	铁
Cobalt	[ˈkəubɔːlt]	Co	27	钴
Nickel	[ˈnɪkl]	Ni	28	镍
Copper	[ˈkɒpə(r)]	Cu	29	铜
Zinc	[zɪŋk]	Zn	30	锌
Gallium	[ˈɡæliəm]	Ga	31	镓
Germanium	[dʒɜːˈmeɪniəm]	Ge	32	锗
Arsenic	[ˈɑːsnɪk]	As	33	砷
Selenium	[səˈliːniəm]	Se	34	硒
Bromine	[ˈbrəumiːn]	Br	35	溴
Krypton	[ˈkrɪptɒn]	Kr	36	氪
Rubidium	[ruːˈbɪdiəm]	Rb	37	铷
Strontium	[ˈstrɒntiəm]	Sr	38	锶

(continued)

Name	Pronunciation	Symbol of the element	Atomic number	Chinese
Yttrium	[ˈɪtrɪəm]	Y	39	钇
Zirconium	[zɜːˈkəunɪəm]	Zr	40	锆
Niobium	[naɪˈəubɪəm]	Nb	41	铌
Molybdenum	[məˈlɪbdənəm]	Mo	42	钼
Technetium	[tekˈniːʃɪəm]	Tc	43	锝
Ruthenium	[ruːˈθiːnɪəm]	Ru	44	钌
Rhodium	[ˈrəudɪəm]	Rh	45	铑
Palladium	[pəˈleɪdɪəm]	Pd	46	钯
Silver	[ˈsɪlvə(r)]	Ag	47	银
Cadmium	[ˈkædmɪəm]	Cd	48	镉
Indium	[ˈɪndɪəm]	In	49	铟
Tin	[tɪn]	Sn	50	锡
Antimony	[ˈæntɪmənɪ]	Sb	51	锑
Tellurium	[teˈljuərɪəm]	Te	52	碲
Iodine	[ˈaɪədiːn]	I	53	碘
Xenon	[ˈzenɒn]	Xe	54	氙
Cesium	[ˈsiːzɪəm]	Cs	55	铯
Barium	[ˈbeərɪəm]	Ba	56	钡
Lanthanum	[ˈlænθənəm]	La	57	镧
Cerium	[ˈsɪərɪəm]	Ce	58	铈
Praseodymium	[ˌpreɪzɪəuˈdɪmɪəm]	Pr	59	镨
Neodymium	[ˌniːəuˈdɪmɪəm]	Nd	60	钕
Promethium	[prəˈmiːθɪəm]	Pm	61	钷
Samarium	[səˈmeərɪəm]	Sm	62	钐
Europium	[juəˈrəupɪəm]	Eu	63	铕
Gadolinium	[ˌgædəˈlɪnɪəm]	Gd	64	钆
Terbium	[ˈtɜːbɪəm]	Tb	65	铽
Dysprosium	[dɪsˈprəuzɪəm]	Dy	66	镝
Holmium	[ˈhəulmɪəm]	Ho	67	钬
Erbium	[ˈɜːbɪəm]	Er	68	铒
Thulium	[ˈθuːlɪəm]	Tm	69	铥
Ytterbium	[ɪˈtɜːbɪəm]	Yb	70	镱
Lutetium	[luːˈtiːʃɪəm]	Lu	71	镥
Hafnium	[ˈhæfnɪəm]	Hf	72	铪
Tantalum	[ˈtæntələm]	Ta	73	钽
Tungsten	[ˈtʌŋstən]	W	74	钨
Rhenium	[ˈriːnɪəm]	Re	75	铼
Osmium	[ˈɒzmɪəm]	Os	76	锇
Iridium	[ɪˈrɪdɪəm]	Ir	77	铱
Platinum	[ˈplætɪnəm]	Pt	78	铂

(continued)

Name	Pronunciation	Symbol of the element	Atomic number	Chinese
Gold	[gəuld]	Au	79	金
Mercury	[ˈmɜːkjəri]	Hg	80	汞
Thallium	[ˈθæliəm]	Tl	81	铊
Lead	[liːd]	Pb	82	铅
Bismuth	[ˈbɪzməθ]	Bi	83	铋
Polonium	[pəˈləuniəm]	Po	84	钋
Astatine	[ˈæstətiːn]	At	85	砹
Radon	[ˈreɪdɒn]	Rn	86	氡
Francium	[ˈfrænsiəm]	Fr	87	钫
Radium	[ˈreɪdiəm]	Ra	88	镭
Actinium	[ækˈtɪniəm]	Ac	89	锕
Thorium	[ˈθɔːriəm]	Th	90	钍
Protactinium	[ˌprəutækˈtɪniəm]	Pa	91	镤
Uranium	[juˈreɪniəm]	U	92	铀
Neptunium	[nepˈtjuːniəm]	Np	93	镎
Plutonium	[pluːˈtəuniəm]	Pu	94	钚
Americium	[ˌæməˈrɪsiəm]	Am	95	镅
Curium	[ˈkjuəriəm]	Cm	96	锔
Berkelium	[bɜːˈkiːliəm]	Bk	97	锫
Californium	[ˌkælɪˈfɔːniəm]	Cf	98	锎
Einsteinium	[aɪnˈstaɪniəm]	Es	99	锿
Fermium	[ˈfɜːmiəm]	Fm	100	镄
Mendelevium	[ˌmendəˈliːviəm]	Md	101	钔
Nobelium	[nəuˈbiːliəm]	No	102	锘
Lawrencium	[lɒˈrensiəm]	Lr	103	铹
Rutherfordium	[ˌrʌðəˈfɔːdiəm]	Rf	104	𬬻
Dubnium	[ˈdʌbniəm]	Db	105	𬭊
Seaborgium	[siːˈbɔːgiəm]	Sg	106	𬭳
Bohrium	[ˈbɔːriəm]	Bh	107	𬭛
Hassium	[ˈhæsiəm]	Hs	108	𬭶
Meitnerium	[maɪtˈnɪəriəm]	Mt	109	鿏
Darmstadtium	[ˈdɑːmʃtætiəm]	Ds	110	𫟼
Roentgenium	[ˌrɒntˈgiːniəm]	Rg	111	𬬭
Copernicium	[kəpɜːˈnɪsiəm]	Cn	112	鎶
Nihonium	[nɪˈhəuniəm]	Nh	113	鉨
Flerovium	[flɛˈrɔːvɪəm]	Fl	114	𫓧
Moscovium	[mɒsˈkəuvɪəm]	Mc	115	镆
Livermorium	[lɪvəˈmɔːrɪəm]	Lv	116	𫟷
Tennessine	[ˈtɛnɪˌsiːn]	Ts	117	鿬
Oganesson	[ɒgəˈnɛsɒn]	Og	118	鿫

A2   General Stems and Affixes

PREFIXES

a-, an-	not, without, lacking
ab-	away, from
ambi-, amphi-	both, around
ana-	up, back, again
ante-	before
anti-	against, opposite
apo-	away from
auto-	self
be-	intensive, to make
bene-	well
bi-	two
by-	aside or apart from the common, secondary
cata-	down
cent-	hundred
circum-	around
co-, com-, con-, col-, cor-	together, with
counter-, contra-	against, opposite
de-	away, down from, negative
deca-, deka-, dec-	ten
demi-	half
di-	two
di-, dif-, dis-	away, negative
dia-	through, across
du-	two
epi-	upon, over, outer
en-, em-	in, into, to make
eu-	good, well
ex-, e-, ec-	out, external
extra-	outside, outward
fore-	before
hemi-	half
hepta-	seven
hetero-	other
hexa-	six
homo-	same
hydr-	water, liquid

hyper-	over, above, beyond
hypo-	under, beneath, down
il-, im-, in-, ir-	not, in, into, on
inter-	between
intra-, intro-	inside, within
macro-	large
mal-	bad, badly
meta-, met-	change
micro-	small
mid-	middle
mini-	small, little
mis-	wrong, unfavorable, bad, badly, hate
mono-	one, alone
multi-	many
neo-	new
non-	not
nona-	nine
oct-	eight
omni-	all
out-	out
over-	above, too much
pan-	all
para-, par-	beside
penta-	five
peri-	around
poly-	many
post-	after
pre-	before
prim-, prin-	first
pseudo-	false
quad-, quart-	four
quint-	five
re-, retro-	backward, back, behind, again
semi-	half
sext-	six
sept-	seven
solo-	alone
sub-, suc-, suf-, sug-, sup-, sus-	under
super-, supra-, sur-	over, above
sym-, syn-, syl-, sys-	with, together
tetra-	four
thermo-	heat

trans-, tra-	across
tri-	three
ultra-	beyond, excessive, extreme
un-	not
under-	beneath
uni-	one

SUFFIXES

-able, -ible, -ble	(*adj.*) capable of, suitable for, fit for
-age	(*n.*) state, quality, act
-al	(*adj.*) related to, like
-al	(*n.*) act, one who, that which
-an	(*n.*) one who
-an, -ian, -ean	(*adj.*) like, related to
-ance, -ence	(*n.*) state, quality, act
-ancy, -ency	(*n.*) state, quality, act
-ant, -ent	(*n.*) one who, that which
-arian	(*n.*) age, section, belief or occupation
	(*adj.*) having the quality of
-ate	(*adj.*) having the quality of
	(*v.*) to make
-crat	(*n.*) a person connect with
-dom	(*n.*) state, quality, act
-ed	(*adj.*) adjective form of nouns and verbs
-ee	(*n.*) one who (passive)
-eer	(*n.*) one who
-en	(*v.*) make
-en	(*adj.*) having the quality of
-er (-or)	(*n.*) one who, that which
-ery	(*n.*) state, quality, act, place of activity
-ese	(*adj. & n.*) like, related to
-ess	(*n.*) feminine
-ferrous	(*adj.*) bearing, causing
-ful	(*adj.*) having the quality of
-fy	(*v.*) to make
-hood	(*n.*) state, quality
-ia	(*n.*) disease, state, quality
-ic (-ical)	(*adj.*) like, related to, having the nature of
-ify	(*v.*) cause, to become
-ing	(*n.*) turns countable nouns into uncountable nouns indicating material
-ion, -tion	(*n.*) state, quality, act

-ish	(*adj.*) like, related to
-ism	(*n.*) state, quality, act, doctrine, system, condition
-ist	(*n.*) member of a party, occupation, etc.
-ite	(*adj.*) related to, having the quality of
-ity	(*n.*) state, quality, act
-ive	(*adj.*) tending to, having the quality of
-ize, -ise, -yze, -yse	(*v.*) cause to be or have, to make
-less	(*adj.*) without
-let	(*n.*) small, unimportant thing
-like	(*adj.*) having the quality of
-logy	(*n.*) discourse, study
-ly	(*adj.*) having the quality of
-ly	(*adv.*) in the manner of …
-ment	(*n.*) state, quality, act
-ness	(*n.*) state, quality
-oid	(*adj.*) like, resembling
-ory, -orium	(*n.*) place where
-ous, -eous, -ious, -ose	(*adj.*) having the quality of, full of
-ship	(*n.*) state, quality
-some	(*adj.*) having the quality of
-ster	(*n.*) a person making or doing something, a member of
-tion, -ation	(*n.*) condition, the act of
-ulent	(*adj.*) quantity of
-ward (-wards)	(*adv.*) toward
-wise	(*adv. & adj.*) in the manner of, as far as…concerned, direction
-y	(*n.*) state, quality, act
	(*adj.*) full of, covered with, having the quality of
	(*n.*) tiny, affectionate name

STEMS

ag, act	do, drive
agr	field
alter, al	other
am	love
ambul	walk
anim	spirit, mind, life
andr	male
ann, enn	year
apt	adjust, fit
aqua	water
arch	first, chief, rule
anthrop	man

aster, astro, stellar	star
audi, audit	hear
aug, auc, aut	increase
auri	ear
best	animal
bibli	book
bio	life
blanc	white
bon, bono, bene	good, well
botan	plant
brev, brief	short
brut	animal
cad, cid, cas	fall, happen
capit	head, main
cap, cip, cept, ceive, cup	take, hold, capture
card	heart
cart, chart	paper
cede, ceed, cess	go, move, yield
cent	hundred
center, centr	center, middle
cephal	head
cert	perceive, make certain
chiro	hand
chrom	color
chron	time
cide, cis	kill, cut
circ, cycl	circle
claim, clam	shout, cry out
clos, clud, claus, clus	shut, close
cogn	know
cor, cord	heart
corp	body
cosm	universe, world
crat	rule
cred	believe
cruc	cross
cur	care
cur, curs, course	run
cult	grow, develop
cycl	circle
demo	people
dens	thick

dent	tooth
derm	skin
di	day
dict, dic	say, speak, tell
doc, doct	teach
domin	rule, master, tyranny
don	give
dorm	sleep
dors	behind
drom	run
duc, duct	lead
ed	eat
em, am	take, carry
equ	equal
ethn	race
fabl, fabul	talk, tell
fac, fic, fect, fact, fict, fig	do, make
fem, femin	female
fer	carry, bring, bear
ferv	heat
fid	trust, faith
fig	shape, form, mold
fil	children
fin	end, limit, boundary
flat	fill, swell
flect, flex	bend
flict	strike
flor, flour	flower
flu	flow
foli	leaf
form	form, shape
forc, fort	strong
fract, frag	break
frater	brother
fug	flee
fus	pour, melt
gam	marriage
gen	birth, family, cause, kind, race
geo	earth
glaci	ice
gon	angle
grad, gress	step

gram, graph	write, writing
grat, grac	please, thank
grav	weigh, heavy
greg	flock, group, collect
gyn, gynec	female
habit	live, home
hal	draw into
heli	sun
hes, her	stick
hetero	different, other
hibit	hold, carry
homo	same
horr	fear
hypn	sleep
ign	fire
insul, isol	island
it	walk, go
ject	throw
jud	judge
junct, jug, join	join, marry, mate
juven	young
lat	carry, bear
later	side
laps	slip, fall
latr, later	worship
lect, leg, lig	choose, gather, read, select
leg	law
lev	lift
lif, liv	live
lingu	tongue, language
liqu	fluid
lite, lith	stone
loc	place
log, loqu, locu	word, discourse, speak
luc, lumin	light
lun	moon
magn	great
man, manu	hand
mand	order
mater, matr	mother, womb
med	heal
med	middle

mega	great
memor	memory
men	month
meter, metr, mens	measure
migr	move
min	less, little
mir	surprise
misc, mix	mix
miss, mit	send, let go
mob, mot, mov	move
morph	form
mort	death
mur	wall
mut	change
nas	nose
nat, nasc	birth, race, nation
nect, nex	link
neur	nerve
nom, nomin	name
note	mark
ology	word, speech, discourse, study of
onom, onym	name
oper	work
opt	sight, light
opt	choose
ora	speak
org, urg	work
ori	rise
ortho	straight, correct
paci	peace
par	equal
parl	speak, talk
pater, patr	father
path, pass	feeling, suffering, disease
ped	child
ped, pod	foot
pel, puls	drive, push
pend, pens	hang, weigh, pay
petr	stone
phil	love, like
phob	fear
phon	sound

phot, phos	light
phyt	plant
pict	draw
pisc	fish
plen, plet, ply	fill, full
plic, plex, ply	fold
polis	city
popul	people
pon, pos	place, put
port	carry
pot	drink
press	press
prim	first
priv	personal
psych	mind, breath, life, spirit, soul
pur	clear
pyr	fire
quer, quest, quir, quis	ask, seek
quiet, quies	rest
ras, rad	rub, remove
rect, reg	straight, right
rid, ris	laugh
rupt	break
rur, rus	country
sal	salt
sanguine	blood
sat, satis, satur	enough, full
scend, scens, scent	climb
sci	know
scop	see, look at
scope	instrument for seeing or observing
scrib, script	write
sec, sect, seg	cut
selen	moon
sen	old
sens, sent	feel
sequ, sec, secut	follow
sed, sid	sit, settle
sist	stand, set
sol	sun
sol	one, alone
somn	sleep

soph	wise
sorb, sorpt	suck in
spec, spect, spi	look
sper	hope
spir	breath, life
sta, stat, stit	stand, set
stell	star
strain, strict, string	tighten
tact, tang, tag	touch
tail	cut
tain, ten, tin	hold, keep
tele	far
tempor	time
tend, tens, tent	stretch, go, strive
terg, ters	rub, remove, clean
termin	finish, end, limit, boundary
terr	earth
terr	frighten
test	witness
therm	heat
the	god
tim	fear
tom	cut
ton	sound
tor, tort	twist
tract	drag, draw
trib	give
trop	turn
trud, trus	push
turb	disorder
urb	city
ut, us	use
vac, van	empty
vad, vas	walk, come
val	strong, powerful, well, effective, worth
veget	plant
velop	fold, wrap
ven, vent	come
ver	true
vers, vert	turn
via, vi	way, road
vid, vis	see

vit, viv	life, live
voc, vok	voice, call
vol, volunt	will
volv	roll, turn
vor	eat
wis, wit	know
zo	animal

A3 Common English Terms Used in the Laboratory

Related Laboratory Equipment

absorbent cotton	脱脂棉	oven	烘箱
adapter	接液管	paddle blender	桨式捣碎器
air condenser	空气冷凝管	parafilm wrap	封口膜
analytical balance	分析天平	pestle	研杵，乳钵槌
beaker	烧杯	petri dish	培养皿
cell culture flask	细胞培养瓶	pinchcock	弹簧夹
centrifuge	离心机	pipette	移液管
condenser	冷凝器	plastic squeeze bottle	塑料挤压瓶
crucible	坩埚	platform balance	托盘天平
crucible cover	埚盖	reagent bottle	试剂瓶
crucible tong	坩埚钳	reducing bushing	异径套管
deionized water	去离子水	ring clamp	环形夹
Dewar	杜瓦瓶，杜瓦容器	ring stand	铁架（台）
distilling tube	蒸馏管	round-bottom flask	圆底烧瓶
drop-dispenser	滴管	scissor	剪刀
electric sterilizer	电灭菌器	screw clamp	螺旋夹
Erlenmeyer flask	锥形瓶	separatory funnel	分液漏斗
evaporating dish	蒸发皿	standard taper equipment	标准锥度设备
extension clamp	万能夹，牵引夹	stemless funnel	无颈漏斗
filter flask	抽滤瓶	stirring rod	搅拌棒
Florence flask	佛罗伦萨烧瓶，平底烧瓶	stopcock	龙头，活栓，活塞
fractionating column	精馏柱	stirring bar	搅拌子
funnel	漏斗	suction bulb	洗耳球
furnace	熔炉	syringe	注射器
gas measuring tube	气体测量管，量气管	test tube	试管
Geiser burette	活塞酸式滴定管	test-tube clamp	试管夹
glass filter crucible	玻璃过滤坩埚	test-tube rack	试管架
glass rod	玻棒	Thiele melting point tube	蒂勒熔点管
glove	手套	Thistle tube	蓟头漏斗
goggle	护目镜	transfer pipette	移液管
graduated cylinder	量筒	tripod	三脚架，三脚支撑物
graduated pipette	刻度吸管	tweezer, forcep	镊子，钳子
Griffin beaker	烧杯	ultrasonic cleaner	超声波清洗器
ground joint	磨口接头	volumetric flask	容量瓶
lab coat	实验服	wash bottle	洗瓶

lab marker	实验室标记笔	watch glass	表面皿
long-stem funnel	长颈漏斗	water aspirator pump	吸水泵
mask	口罩	weighing bottle	称量瓶
microscope	显微镜	weighing paper	称量纸
Mohr burette	莫尔滴定管	weighing spatula	称量勺
mortar	研钵		

Related Chemical Reactions and Stoichiometry

disproportionation	歧化	heat	热，加热
neutralization	中和	ignite (burn)	点燃（燃烧）
catalysis	催化	catalyst	催化剂
electrolysis	电解	catalyze	催化
exothermic reaction	放热反应	electrolyze	电解
endothermic reaction	吸热反应	euthermic	增温的，发热的
reversible reaction	可逆反应	exothermic	放热的
forward reaction	正向反应	endothermic	吸热的
reverse reaction	逆向反应	ambient temperature	环境温度
spontaneous reaction	自发反应	under standard pressure	在标准压力下
nonspontaneous reaction	非自发反应	solubility	溶解度
atomic mass / weight	原子质量，原子量	soluble	可溶的
molecular weight	分子量，相对分子质量	insoluble	不溶的
amount (of substance)	（物质的）量	slightly soluble	微溶的
mole	摩尔	very soluble	易溶的
number of moles	摩尔数	precipitate	沉淀
molar mass	摩尔质量	aqueous solution	水溶液
molar volume	摩尔体积	milky	浑浊不清，乳白色的
molarity	摩尔浓度	inflammation	燃烧
limiting agent	限量（反应）试剂	luminous	发光
excess agent	过量（反应）试剂	concentration	浓度
reactant	反应物	yield	产量，产率
product	产品		

References

[1] Klein D R. Organic Chemistry. 4th ed. New York: Wiley, 2020.

[2] Atkins P, De Paula J. Physical Chemistry. 8th ed. New York: W. H. Freeman and Company, 2006.

[3] Harris D C. Quantitative Chemical Analysis. 8th ed. New York: W. H. Freeman and Company, 2010.

[4] Weller M T, Overton T, Rourke J. Inorganic Chemistry. 7th ed. Oxford: Oxford University Press, 2018.

[5] Morss L R, Edelstein N M, Fuger J. The Chemistry of the Actinide and Transactinide Elements. 3rd ed. Dordrecht: Springer, 2006.

[6] Hellwich K H, Hartshorn R M, Yerin A, et al. Brief Guide to the Nomenclature of Organic Chemistry (IUPAC Technical Report). Pure and Applied Chemistry, 2020, 92 (3), 527-539.

[7] Hartshorn R M, Hellwich K H, Yerin A, et al. Brief Guide to the Nomenclature of Inorganic Chemistry. Pure and Applied Chemistry, 2015, 87 (9-10), 1039-1049.

[8] Brown W H, Poon T. Introduction to Organic Chemistry. 6th ed. New York: Wiley, 2016.

[9] Sorenson T S. Surface Chemistry and Electrochemistry of Membranes. Boca Raton: CRC Press, 1999.

[10] Skoog D A, Holler F J, Crouch S R. Principles of Instrumental Analysis. 6th ed. Boston: Cengage Learning, 2007.

[11] Laidler K J. The World of Physical Chemistry. Oxford: Oxford University Press, 1995.

[12] Greenwood N N, Earnshaw A. Chemistry of the Elements. 2nd ed. Oxford: Butterworth-Heinemann, 1997.

[13] Keinan E, Schechter I. Chemistry for the 21st Century. Darmstadt: Wiley-VCH, 2001.

[14] Lewis M. Advanced Chemistry Through Diagrams. 2nd ed. Oxford:Oxford University Press, 2001.

[15] Harrison J. Secretarial Duties. 10th ed. London: Pearson Schools, 1996.

[16] Periodic Table of Elements. https://byjus.com/periodic-table/.

[17] Chemistry from Science, Tech, Math>Science, ThoughtCo. http://www.thoughtco.com/chemistry-4133594.

[18] 中国化学会有机化合物命名审定委员会.有机化合物命名原则. 北京：科学出版社，2018.

[19] 从丛, 李咏燕. 学术交流英语教程. 南京:南京大学出版社, 2003.

[20] 张裕平，王丙星，龚文君. 化学化工专业英语.2 版. 北京：化学工业出版社，2021.

[21] 胡鸣，刘霞. 化学工程与工艺专业英语.2 版. 北京：化学工业出版社，2020.

[22] 魏高原. 化学专业基础英语.7 版. 北京：北京大学出版社，2011.

[23] Michael Lewis. 化学专业英语基础. 荣国斌, 译. 上海：上海外语教育出版社，2000.